ANIMAL BODIES, HUMAN MINDS: APE, DOLPHIN, AND PARROT LANGUAGE SKILLS

DEVELOPMENTS IN PRIMATOLOGY: PROGRESS AND PROSPECTS

Series Editor:

Russell H. Tuttle
University of Chicago, Chicago, Illinois

This peer-reviewed book series will meld the facts of organic diversity with the continuity of the evolutionary process. The volumes in this series will exemplify the diversity of theoretical perspectives and methodological approaches currently employed by primatologists and physical anthropologists. Specific coverage includes: primate behavior in natural habitats and captive settings: primate ecology and conservation; functional morphology and developmental biology of primates; primate systematics; genetic and phenotypic differences among living primates; and paleoprimatology.

ALL APES GREAT AND SMALL
VOLUME 1: AFRICAN APES
Edited by Birute M. F. Galdikas, Nancy Erickson Briggs, Lori K. Sheeran, Gary L. Shapiro and Jane Goodall

THE GUENONS: DIVERSITY AND ADAPTATION IN AFRICAN MONKEYS
Edited by Mary E. Glenn and Marina Cords

ANIMAL BODIES, HUMAN MINDS: APE, DOLPHIN, AND PARROT LANGUAGE SKILLS
William A. Hillix and Duane M. Rumbaugh

COMPARATIVE VERTEBRATE COGNITION: ARE PRIMATES SUPERIOR TO NON-PRIMATES?
Lesley J. Rogers and Gisela Kaplan

ANIMAL BODIES, HUMAN MINDS: APE, DOLPHIN, AND PARROT LANGUAGE SKILLS

William A. Hillix

Professor Emeritus
Department of Psychology
San Diego State University
San Diego, California

And

Duane M. Rumbaugh

Regents Professor Emeritus
Departments of Psychology and Biology
Georgia State University
Atlanta, Georgia
and
Iowa Primate Learning Sanctuary
Des Moines, Iowa

KLUWER ACADEMIC / PLENUM PUBLISHERS
New York, Boston, Dordrecht, London, Moscow

ISBN 0-306-47739-4

©2004 Kluwer Academic/Plenum Publishers, New York
233 Spring Street, New York, New York 10013

http://www.wkap.nl/

10 9 8 7 6 5 4 3 2 1

A C.I.P record for this book is available from the Library of Congress

Permissions for books published in Europe: *permissions@wkap.nl*
Permissions for books published in the United States of America: *permissions@wkap.com*

Printed in the United States of America

PREFACE

Several books chronicle attempts, most of them during the last 40 years, to teach animals to communicate with people in a human-designed language. These books have typically treated only one or two species, or even one or a few research projects. We have provided a more encompassing view of this field. We also want to reinforce what other authors, for example, Jane Goodall, Sue Savage-Rumbaugh, Penny Patterson, Birute Galdikas, and Roger and Deborah Fouts, so passionately convey about our responsibility for our closest animal kin.

We have very different backgrounds. I (William A. Hillix) have written about the history of psychology and the systems and theories that are part of that history. Duane M. Rumbaugh, on the other hand, has contributed greatly to the animal language studies that have become an increasingly important part of psychology within recent years. Duane began his work with primates at San Diego State University in collaboration with the San Diego Zoo, and continued at Yerkes Laboratory in Atlanta, Emory University, and Georgia State University, where he and Sue Savage-Rumbaugh co-established the Language Research Center in Decatur, Georgia. He served as director of the Center for many years. His pioneering work with Lana and other chimpanzees established a tradition that continues to the present day. He used lexigrams as the medium of communication with chimpanzees, and later, with Sue Savage-Rumbaugh, moved toward an emphasis on the language comprehension of bonobos and chimpanzees rather than on the language production that had received most of the early attention. In recognition of the pioneering work of Japanese primatologists, we have asked one of their contemporary researchers, Dr Tetsuro Matsuzawa of Kyoto University, to describe the research in Japan that bears upon the learning and use of symbols in language and perception. We hope that this book will lead to a greater appreciation of animals and of our very closest genetic relatives—other human beings. If we can see ourselves in the mirror of chimpanzees, who are only about 98.7 percent genetically identical to us, why do we have such difficulty in seeing ourselves in the mirror of other humans, those very nearly genetically identical animals who differ from us mostly in that they may speak another language or follow another religion? It may be too much to believe that this book could contribute to a greater appreciation of human—as well as of animal—life, but that is nevertheless our hope.

In this book we will survey what was known or believed about animal language throughout history and prehistory, and summarize current knowledge and the controversy that swirls around it. We will identify, and try to settle, most of the problems in interpreting the animal behaviors that have been observed in studies of animal language ability. We will also present our best guesses about where research in animal language will go from here.

We wish to thank the many researchers whose work we have cited and the animals that made their research possible. They are named in our Cast of Characters of contributors to animal language research (our apologies to both the animals and the people whom we have neglected). Duane M. Rumbaugh wishes to acknowledge long-term support from the National Institute of Child Health and Human Development (HD06016 and 38051). Special thanks are due to Roger and Deborah Fouts, Allen Gardner, Louis Herman, Penny Patterson, Irene Pepperberg, David Premack, and Sue Savage-Rumbaugh, all of whom read and commented on the chapters describing their work. We also thank them and Lyn Miles for providing and allowing us to print the photographs that enliven the text. Any errors remaining in this book are, of course, our responsibility. Every researcher and participant has taught us a great deal about humans, animals, and the degree to which language can help to unite them.

William A. Hillix
Duane M. Rumbaugh

CONTENTS

ANIMAL BODIES, HUMAN MINDS: APE, DOLPHIN, AND PARROT LANGUAGE SKILLS

CHAPTER ONE

A Chronology of Events in Animal Language Research

ou may wish to skim this chronology to get a rapid overview of how research on trying to teach animals a human language has progressed. Interest in animal language is ancient, but its scientific study is strikingly modern.

c.1000 BC	The author(s) of the *Book of Genesis* reported a conversation between Eve and the serpent.
c.600 BC	Aesop wrote his fables about talking animals.
c.500 AD	The scribe Horapollo Nilous reports that Egyptian priests believed that sacred baboons had the power of reading and writing, and therefore gave baboons newly arrived in the temple a test of their language abilities.
c.1280	Approximate date of birth of William of Occam (also spelled Ockham), who is usually credited with formulating the principle of parsimony; this principle is used to eliminate unnecessary speculations from scientific discussions, including those of animal language.
1349	William of Occam dies.
1633	Samuel Pepys born.
1661	Samuel Pepys sees a "great baboone," probably a chimpanzee, and in his diary of August 21 suggests that "it might be taught to speak or make signs."
1703	Samuel Pepys dies.
1709	French philosopher, Julien Offray de la Mettrie, born.
1751	La Mettrie dies.
1838	Wilhelm Von Osten, first owner of horse Clever Hans, born.
1848	Richard Lynch Garner, early student of great apes, born.

Carl Stumpf, member of September Commission, which studied abilities of Clever Hans (see below) born.

George Romanes, writer on animal intelligence, born.

1852   C. Lloyd Morgan, famous for his canon stating preference for simplicity in explaining animal abilities, born.

1859   Charles Darwin publishes *The origin of species*.

1861   Karl Krall, owner of horses with unusual abilities, born.

1866   Von Osten moves to Berlin.

William Henry Furness 3rd born.

1874   Oskar Pfungst, who later explained Clever Hans' performance as depending on cues from observers, born.

1876   Robert Yerkes, student of primates and founder of the first primate field station in the United States, born.

1886   Hugh Lofting, of Dr Dolittle fame, born.

1890   Von Osten works with Hans I.

1891   C. Lloyd Morgan publishes *Animal life and intelligence*.

Othello, a gorilla later purchased by Richard Lynch Garner, born.

1892   Garner arrives in Africa, goes into cage in Gabon.

Moses, Garner's favorite chimpanzee, born in Africa.

Aaron, a kulu-kamba who was a favorite of Garner, born.

Othello dies; Garner publishes *The speech of monkeys*.

1894   George Romanes dies.

1895   Romanes' book, *Animal intelligence*, published.

1896   Garner publishes *Gorillas and chimpanzees*.

1898   Winthrop Kellogg born.

1900   First public appearance of the horse, Clever Hans.

1903?   Signor Emilio Rendich gave what later appeared to be the correct explanation of Hans' abilities; demonstrated that his dog Nora could learn to do what Hans did, using cues from Rendich.

1904   Clever Hans gives demonstrations in Berlin. September Commission investigates Hans' abilities.

B. F. Skinner, author of *Verbal behavior*, born.

1907   Pfungst publishes *Clever Hans, the horse of Mr. von Osten*.

1908   Unknown Soldier, orangutan not named by Furness, born; the first orangutan known to utter a word of human language.

1909   Von Osten dies.

Karl Krall gets Clever Hans.

1912   Krall publishes *Denkende Tiere (Thinking animals)*.

Furness' orangutan, Unknown Soldier, dies.

1913   Max Rothmann, who established the Anthropoid Field Station of the Prussian Academy of Sciences in the Canary Islands, and Eugen Tueber, its first director, suggest teaching a gestural language to chimpanzees.

Nadesha Kohts adopts the 1½-year-old chimpanzee, Joni.

1915   John Lilly born.

1916   William Furness III suggests teaching chimpanzees a gestural language; reports that he taught a female orangutan to say "papa," "cup," and "th."

Nadesha Kohts stops working with Joni.

William Lemmon born.

1917   Franz Kafka writes "A report to the academy," a story about a talking chimpanzee.

Nadesha Kohts' chimpanzee, Joni, dies.

1920   Hugh Lofting publishes *The story of Dr Dolittle*.

Richard Lynch Garner dies.

William Henry Furness 3rd dies.

Thomas A. Sebeok, severe critic of animal language studies, born.

1925   Roger Brown born.

David Premack born.

Robert Yerkes suggests in the book *Almost human* that great apes might be taught sign language.

Nadesha Kohts' son, Roody, born; Kohts compared his development to that of the chimpanzee, Joni, born about 13 years earlier.

1928   Warden and Warner describe German Shepherd dog Fellow's comprehension of English commands.

Noam Chomsky born.

1929   Duane Rumbaugh born.

Ann Premack born.

Karl Krall dies.

1930   R. Allen Gardner born.

Chimpanzee Gua born in Cuba on November 15.

Donald Kellogg born on August 31; raised with Gua for 9 months.

1931   Gua, aged 7½ months, adopted by W. N. Kellogg and L. A. Kellogg.

1932   Ronald Schusterman, student of sea lion abilities, born.

Oskar Pfungst dies.

Gua returned to chimpanzee colony.

1933   Beatrix Gardner born.

The Kelloggs publish *The ape and the child*, summarizing their work with Gua and Donald.

1934	Philip Lieberman born.
1935	Nadesha Kohts publishes her observations of Joni and Roody.
1936	Karl Stumpf dies.
	C. Lloyd Morgan dies.
	Herbert Terrace born.
1943	Roger and Deborah Fouts both born.
	Japanese researcher, Kojima, born.
	Ronald Cohn, Koko's photographer and playmate, born.
1944	Lyn Miles, researcher with the orangutan Chantek, born.
1946	Sue Savage-Rumbaugh born.
1947	Chimpanzee Viki born in September; Cathy and Keith Hayes take her into their home and raise her almost as a human child.
	Francine Patterson, teacher of Koko, born.
1948	Hugh Lofting dies.
1949	Sarah Boysen born.
	Irene Pepperberg born.
1951	Cathy Hayes publishes *The ape in our house*, a report of the work she and her husband, Keith, had done with Viki.
1954	Viki dies of encephalitis.
	Steven Pinker, critic of ape language studies, born.
1956	Robert Yerkes dies.
1957	B. F. Skinner publishes *Verbal behavior*.
1959	Chomsky excoriates Skinner for his ideas on language expressed in his book, *Verbal behavior*.
1961	David Washburn, student of Duane Rumbaugh and current Director of the Language Research Center, born.
1964	Lucy born in Oklahoma; taken into the Temerlin household.
	Premacks try to teach chimpanzees to use a joystick controlling a phoneme generator that produces human-like sounds.
1965	Chimpanzee Washoe born in West Africa.
1966	Beatrix and Allen Gardner start to teach Washoe sign language, 305 years after Pepys suggested the possibility.
1967	Booee born in Bethesda, Maryland.
1968	Ann and David Premack start to teach their chimpanzee, Sarah, to communicate by writing with plastic symbols.
	Bruno born at Lemmons' Institute.
1969	Allen and Beatrix Gardner report in *Science* that they have taught 85 signs resembling those of American Sign Language to Washoe.
	Ally born.

1970   Lana born; later taught to use first lexigram board under the direction of Duane Rumbaugh.

Language Analog (LANA) project begun by Rumbaugh.

Bonobo Matata, to become Kanzi's stepmother, born in the wild.

Washoe moves to Oklahoma with Roger Fouts.

1971   Francine Patterson hears Gardners speak; Gorilla Koko (full name Hanabi Ko for "fireworks child") born on July 4; later taught to sign by Patterson.

1972   Chimpanzee Moja, full name Moja LEMSIP, born on November 18 at the LEMSIP lab in New York; she arrives at Gardners' laboratory in Reno the following day. (Moja is Swahili for "first.")

Winthrop Kellogg dies.

Francine Patterson starts work with Koko.

1973   Chimpanzee Pili born in Georgia; he joins Moja in Reno 2 days later.

Chimpanzee Nim Chimsky born in Oklahoma and loaned to Herbert Terrace. He becomes a center of controversy about the ability of chimpanzees to produce language.

Chimpanzee Sherman born in Georgia; he and chimpanzee Austin, under the tutelage of Duane and Sue Savage-Rumbaugh, became the first chimpanzees to communicate with each other via lexigrams.

Gorilla Michael born; became companion of Koko under tutelage of Patterson.

1974   Austin born in Georgia.

1975   Chimpanzee Tatu born on December 30 in Oklahoma.

Chimpanzee Pili dies of leukemia.

1976   Tatu arrives at Gardners' laboratory on January 2.

Michael joins Koko in Patterson's research project when he is about 3½ years old.

Chimpanzee Dar born in New Mexico on August 2. He joins Moja and Tatu at Gardners' laboratory on August 6.

Washoe gives birth to infant who survives for only 4 hours.

Alex, the African Gray parrot, born.

David Premack publishes *Intelligence in ape and man*.

1977   Lucy returns to the wilds in Gambia with Janis Carter.

Chantek, the orangutan taught to sign by Lyn Miles, born.

Rumbaugh publishes *Language learning by a chimpanzee: The Lana project*.

1978   Chimpanzee Loulis, born in Georgia, later becomes stepson of Washoe in Oklahoma.

1979   Terrace et al. report largely negative results with Nim Chimsky.
            Washoe gives birth to Sequoyah in January; he dies in March and is
            replaced by Loulis 15 days later.
            Moja, then 7, joins Washoe and Loulis in Oklahoma.
1980   Bonobo Kanzi born on October 28 at Yerkes Research Center in
            Georgia; immediately stolen from his mother, Lorel, by Matata, who
            already had another infant, Akili.
            Thomas Sebeok and Jean Umiker-Sebeok publish a book, based on a
            conference, that severely criticizes ape language studies.
            Washoe, Loulis, and Moja move to Central Washington University with
            Roger and Deborah Fouts.
1981   Tatu and Dar, both now 5-year-olds, join the Fouts group in Washington.
1982   Ristau and Robbins publish a balanced summary and critique of ape
            language research.
1983   Bonobo Mulika born.
1984   Philip Lieberman says that chimpanzees' vocal tracts could generate
            complex language "if other factors are present."
1985   Chimpanzee Panzee, bonobos Panbanisha and, probably, bonobo
            P-Suke born.
1986   Alia, daughter of Jeannine Murphy, born; her language development
            was compared to Kanzi's, beginning when she was 2 and he was 7½.
            William Lemmon dies.
            Sue Savage-Rumbaugh publishes *Ape language: from conditioned response
            to symbol*, primarily based on work with chimpanzees Sherman and Austin.
1987   Bonobo Tamuli born.
1988   Lucy found dead on Baboon Island in The Gambia, probably murdered
            by poachers.
1990   B. F. Skinner dies.
            Book entitled *Congo* by Michael Crichton stars a talking gorilla.
1993   Report compares comprehension of bonobo, Kanzi, with that of human
            child, Alia.
1994   Sue Savage-Rumbaugh and Roger Lewin publish *Kanzi*.
1995   Beatrix Gardner dies on June 8.
            Movie *Congo* based on Crichton's book is released.
1996   Chimpanzee Austin dies of unknown causes.
1997   Roger Fouts publishes *Next of kin*.
1998   Nyota, son of bonobos Panbanisha and P-Suke, born at Georgia State's
            Language Research Center. Sue Savage-Rumbaugh and her students
            provide cross-cultural experiences for him.

2000   Gorilla Michael dies on April 19 at the age of 27, apparently of fibrosing cardiomyopathy, leading to heart failure.

Bonobo Tamuli dies of a congenital heart defect.

2001   John Lilly dies at 86, having written 19 books, including *Man and dolphin* and *The mind of the dolphin*. Two movies based on his work were *Altered states* and *The day of the dolphin*.

Thomas Sebeok dies at 81 on December 21.

2002   Chimpanzee Moja dies on June 6 of an entrapped bowel.

CHAPTER TWO

An Overview of Animal Language

CONVERSATIONS WITH ANIMALS

On a recent day in Decatur, Georgia, the following conversation took place:

Panbanisha: Milk, sugar.
Liz: No, Panbanisha, I'd get in a lot of trouble if I gave you milk with sugar.
Panbanisha: Give milk, sugar.
Liz: No, Panbanisha, I'd get in a lot of trouble.
Panbanisha: Want milk, sugar.
Liz: No, Panbanisha, I'd get in so much trouble. Here's some milk.
Panbanisha: Milk, sugar. Secret.

Panbanisha is a bonobo. (Bonobos used to be called pygmy chimpanzees, but they are a separate species whose formal designation is *Pan paniscus.* To contrast them with pygmy chimpanzees, chimpanzees used to be called common chimpanzees, but that is no longer either necessary or correct.)

Liz Pugh is one of Panbanisha's human caretakers. Panbanisha was talking by pointing to symbols on a lexigram board. Each symbol stands for a single word. Liz was talking back in English.

Do you think Panbanisha understood this exchange? What would it take to convince you? Did Panbanisha really know what it meant to keep a secret?

Panbanisha is one of several animals, both bonobos and chimpanzees, living at the Language Research Center (LRC) in Decatur, Georgia. Drs. Duane Rumbaugh and Sue Savage-Rumbaugh have worked intensively with these animals for over 25 years, giving them opportunities to socialize with humans and other animals and to learn the meanings of symbols on special boards that, in some cases, generate English words via computer speech production.

The goal of the research is to explore the limits of ape comprehension and production of a human-designed language.

Today Kanzi, a 22-year-old bonobo at the LRC, understands more human language than any other nonhuman animal in the world. When Sue Savage-Rumbaugh first said to Kanzi, "Can you make the doggie bite the snake?" he had never heard the sentence before. Nevertheless, he searched among the objects present until he found a toy dog and a toy snake, put the snake in the dog's mouth, and used his thumb and finger to close the dog's mouth on the snake. In rigorous testing, Kanzi, beginning when he was 7½ years old, responded correctly to 74 percent of over 600 unique sentences similar to the doggie sentence. An intelligent 2-year-old human female child, Alia, with whom Kanzi was compared, responded correctly to 65 percent of the same sentences.[1] Both were tested over a period of 9 months, so Alia was over 2½ at the end of testing. Kanzi's performance was an amazing demonstration of an animal's ability to understand human language.

WHAT ANIMAL LANGUAGE RESEARCHERS ARE ASKING

Early researchers were most interested in the question "Can apes talk?" The answer was soon clear: either they could never talk, or it was going to be forbiddingly difficult to teach them.[2]

It is tempting to conclude that, even if apes have some language ability, it will never be expressible via the vocal channel. Researchers with marine mammals have made no attempt, so far as we know, to have dolphins or sea lions communicate via the vocal channel, although a dolphin can imitate computer-generated sounds. At present the only animals making extensive use of the vocal channel are grey parrots, with the parrot named Alex by far the most adept among them. However, Sue Savage-Rumbaugh believes that her bonobos, Kanzi and Panbanisha, are developing mixed English-bonobo vocalizations.

Nadesha Kohts reported[3] that her chimpanzee, Joni, had absolutely no tendency to imitate human speech, and concluded that chimpanzees would never speak. The Kelloggs[4] were disappointed to find that their chimpanzee, Gua, also had no tendency to imitate human speech, even though she imitated motor actions.

Joni was about 1½ years old when Kohts obtained him, and Gua was about 7½ months old when the Kelloggs adopted her temporarily. Both might have been beyond the age of optimal chimpanzee language learning, but Viki, another chimpanzee, was raised from infancy by Keith and Cathy Hayes. Viki occasionally babbled at first, and each day, upon coming home from work, Keith would ask Cathy what Viki "said" that day. The Hayeses were bitterly disappointed when Viki stopped babbling at the age of 4 months, and later could be taught only three (or possibly four) words, despite the Hayeses' persistent application of systematic teaching techniques.

Because of these difficulties, researchers are trying to find other ways to communicate with animals. Better methods are needed if the mental lives of animals are to be compared to the mental lives of humans. Just how close are we humans to each of the other species studied, in terms of language and abilities related to language? What problems can animals solve without language, and can they solve them better after they have had language training? How can the lessons learned in teaching animals be applied to human children? How, and how much, do animals' abilities to learn language differ from humans' and from other species'? Are the abilities that underlie language general or highly specialized abilities? The answers to these fascinating questions are among those that motivate today's researchers.

WHY IS IT SO HARD FOR CHIMPANZEES TO TALK?

People often claim that chimpanzees cannot be taught to talk because they lack a human vocal tract. Certainly their vocal tracts differ from those of humans, but that may not be why chimpanzees have so far been unable to produce vocal language. It may not be the case forever that chimpanzees cannot talk; all we know for sure is that efforts to date have failed. The vocal tract of the gray parrot is much more different from the human vocal tract than is the chimpanzee's, but Alex, the gray parrot, has a spoken vocabulary of about 90 English words.

How do chimpanzees' vocal tracts differ from humans'? For one thing, chimpanzees' vocal cords, like human infants', are placed higher in the vocal tract. The entrance to the trachea in human infants and chimpanzees can be closed off so that breathing and swallowing can occur at the same time without choking. As humans develop, the larynx descends to the position marked by the voice box, or Adam's apple, in the throat, and we do choke if we try to talk and swallow at the same time. The adult human arrangement must have positive adaptive value because it makes speech more facile, since it has the obvious disadvantage that it can lead to death by choking. Thus at least some ability to speak must have preceded the descent of the larynx, because otherwise there would have been no adaptive advantage to encourage its descent.

The chimpanzee tongue is longer and has less room than the human tongue to move vertically; that creates a problem in producing the full range of vowels that are found in human languages.[5,6] The shape of the vocal tract is different in the two species, which also complicates the speech production problem for the chimpanzee. Once the larynx has descended, the adult human has two vocal tubes of approximately equal length, which is necessary to produce good versions of the vowels i and u. The chimpanzee, with one tube longer than the other, cannot produce good versions of these vowels.

However, computer simulation work conducted by Philip Lieberman[5] indicated that chimpanzees' vocal tracts could produce many more human-like sounds than they do produce. And people with extraordinary defects in the vocal tract manage to speak anyway; children without a tongue or without a larynx have nevertheless learned to speak, sometimes rather quickly and clearly![7]

Chimpanzees probably lack the neural control—that is, the brain centers and the neural connections—that can fine-tune the vocal apparatus, rather than being limited primarily by the vocal tract in producing human-like language. An analysis of vocal development in nonhuman primates[8] may shed some light on the vocal shortcomings of chimpanzees, as well as of other primate species. Seyfarth and Cheney distinguish between three aspects of vocal development: the production of sounds, learning to produce them in the correct contexts, and responding correctly to the sounds of others. Their examination of the literature on these subjects indicates that sound production is largely innate, with infant calls from a surprisingly early age sounding much like the calls of adults. However, more flexibility is available to respond to environmental influences that teach animals to use their calls in, and only in, appropriate circumstances. And there is perhaps even more dependence on experience in learning to respond appropriately to the vocalizations of others. If Seyfarth and Cheney are correct, it would be easier to modify comprehension than to modify production for nonhuman primates. Vocal production may depend so much on genetic programming that it is difficult or impossible to modify it sufficiently to support a human-designed language. Thus the best evidence is that the limitation in nonhuman primate vocalization, contrary to what one often hears, lies in a lack of neural connectivity and plasticity rather than in a differently configured vocal tract. An analysis of the evolutionary history of the human speech organs[9] is consistent with this position. Wind believes that there are no differences in mammalian vocal tracts that are fundamental enough to preclude vocal languages, and thus that the difficulties are primarily in cerebral development.

Candland,[10] however, rejects what he sees as the above party line on why chimpanzees can not talk. "The prominent view that chimpanzees could never learn to articulate because they lacked the necessary assemblies of cells in the brain…The area of the brain implicated in human speech is lacking in the chimpanzee brain; therefore (goes the faulty logic), chimpanzees cannot have speech" (p. 284). Candland probably objects to this logic because chimpanzees might control the vocal tract by using different brain areas from those used by humans; it may well be that brain cells are flexible as to their functions, and a particular area could perform several different functions, depending on environmental circumstances. For example, the visual areas in blind humans may come to function as analyzers of acoustic or tactile information. It is also possible that the fact on which the negative argument is based is *not* a fact, and that chimpanzees have areas homologous to all of the speech areas of the human brain, some of which are reduced in size or found in different locations.

Some writers emphasize the close connection between chimpanzee emotion and chimpanzee vocalization.[11] Without the ability to vocalize voluntarily, human communication via language would be extremely limited. The Gardners, whose chimpanzee, Washoe, was the first to be taught a large number of signs, considered the above possible difficulties with chimpanzee vocalization, and preferred the "emotional" explanation:[11]

> [It has been] proposed that chimpanzees cannot learn English because they cannot form the phonemes of human speech; that the impediment is in the design of their vocal tract. But human beings can speak intelligibly even when they must overcome severe injury to their vocal tract. In our view, it is the obligatory attachment of vocal behavior to emotional state that makes it so difficult, perhaps impossible, for chimpanzees to speak English words. They can, however, use their hands in the arbitrary connections between signs and referents (p. 49).

The Gardners thus emphasize the inability to bring the speech apparatus under voluntary control. An example is that chimpanzees have great difficulty in suppressing food barks in the presence of food. Jane Goodall observed one chimpanzee who barked repeatedly upon finding food. Unfortunately, a more dominant chimpanzee always heard the bark and, as a result, came running and stole the food. The unfortunate barker finally learned to suppress the bark, and pretended not to see a new food source until the dominant chimp was out of sight. However, the dominant chimp had hidden behind a bush, and promptly came back to steal the food the moment the submissive chimp picked it up.

The Kelloggs' observations of Gua[4] are consistent with the Gardners' beliefs about the compulsory connection between emotion and vocalization: "On the whole, it may be said that she never vocalized without some definite provocation, that is, without a clearly discernible external stimulus or cause. And in most cases this stimulus was obviously of an emotional character" (p. 281). Catherine Hayes' offhand comments[12] also support this line of reasoning; she said that Viki had not made a *voluntary* sound for months before she began language training. The implication was that Viki had made many involuntary chimpanzee sounds in the interim between her last playful babble and the beginning of training.

The bonobos Kanzi and Panbanisha are exceptions to the above rules. They both try to imitate human speech and can vocalize voluntarily. Kanzi even produced a quite recognizable rendition of "right now," and Sue Savage-Rumbaugh sometimes recognizes other utterances. Kanzi sometimes expresses frustration about difficulty in imitating by pointing to his mouth or throat when he cannot make the sound that he wants to make!

Roger Fouts made a fascinating observation[13] that might be relevant to the extraordinary difficulty that great apes have with vocal communication. He taught sign language to uncommunicative autistic children. A few weeks after they started sign language training they spontaneously spoke their first words! Fouts' explanation is that the same kind of motor dexterity is needed for hand movements and the movements associated with vocalization. Once the necessary dexterity is achieved in one venue, it generalizes to the other. This is consistent with the belief that the areas that control other motor functions overlap the areas that control the speech apparatus.

There are alternative explanations for the results that Fouts achieved; for example, it might be that the autistic children finally caught on to the communicative possibilities that had previously been hidden from them. However, reports of other ways of teaching uncommunicative children have not mentioned the spontaneous emergence of vocal language, which lends credence to Fouts' hypothesis. The suggestion for researchers with the great apes is that they give young apes training in manual dexterity at a very early age and try to encourage its generalization to vocal dexterity, and for researchers with uncommunicative children that they continue to examine Fouts' hypothesis. That could produce improvements in vocalization for all great apes, including humans.

SPEECH IN NON-PRIMATES

If human speech is so difficult to produce for our closest genetic neighbor, it should be much more difficult for animals more biologically different from us—the horse, the elephant, the porpoise, the seal, the dog, and the parrot. Unfortunately for a genetic explanation, this difficulty is much less for the parrot, which is genetically farthest from us among those animals. (And the parrot's vocal apparatus is quite unlike ours; the parrot has a double trachea and a double larynx (the bird version of our larynx is called a syrinx). As everyone knows, parrots use their unusual vocal apparatus to produce remarkable imitations of human speech. Their speech, however, does not *look* as much like human speech when a spectrographic analysis is displayed as it sounds like speech to the human ear.[14] With the other animals, it seemed obvious from the beginning that they would not be able to speak as humans do; some humans can produce remarkable imitations of horses and pigs, but the reverse is not true (at least not yet)! Some dogs have been alleged to make recognizably human sounds; if these claims are true, they constitute another surprise from the point of view of genetic (and vocal tract) relatedness.

Further research is necessary to find out whether ingenious future investigators can design early intensive training in vocalization to overcome the barriers that have so far prevented non-human primates from communicating to any significant degree in human vocal language. Ristau and Robbins[15] take an optimistic view, which the reader will later see has already been vindicated with respect to chimpanzee comprehension. "It may also be that the potential limit of chimpanzees' accomplishments has not been reached in any current ape language project, and awaits the successful meeting of a brilliant chimpanzee and a training procedure refined by experienced researchers" (p. 202). Perhaps the suggestion above that manual dexterity be taught first will be a key component!

Despite the barriers, chimpanzees should be able to make enough sounds to produce a complex language. In principle, that is almost trivially true; Morse code, which has only dots and dashes as its basic sounds, can be used to produce "a complex language." Computers, too, have only two fundamental states for each element, and they perform complex operations on language.

There is a tradeoff between the number of different sounds (or of other units like signs) that can be produced and the number required to transmit a given message. For example, if you can produce 32 different sounds, you can send 32 different messages by making just one sound; but if you can produce only two different sounds, you will have to use five of them to tell the listener which of the 32 messages you are communicating. The number of phonemes used in most human languages is somewhat more than 32, so the example is realistic. The tradeoff between the number of distinct sounds and the number of separate sounds needed to convey a message is one explanation for the fact that the use of words as units has been more effective than the use of letters as units in animal language studies. It apparently is much easier for an animal to learn that a single symbol refers to an apple than to comprehend that five letters, each of which can be used in many non-apple combinations, stand for an apple. Humans also learn words before they learn how to spell.

BYPASSING THE VOCAL CHANNEL

In the book[2] he published at the end of the 19th century, Richard Lynch Garner included a picture of an infant chimpanzee in front of an apparatus containing alphabet blocks on rods that could be turned to select letters; turning several rods to the right position could spell out words. Today we find it difficult to imagine a technique less likely to succeed; we can see a puzzled look on the poor chimpanzee's face.

However, there are two lessons to be drawn from this picture. First, Garner was obviously frustrated with his attempts to get apes to use vocal language. Second, he was already trying to circumvent the difficulty.

Researchers after Garner did not immediately give up on teaching the apparently human-like chimpanzee vocal language. Garner himself tried hard to teach vocal language, and reported (there was apparently no opportunity for independent observers to confirm the report) that one of his animals learned the word "feu," the French equivalent of the English word "fire".

With our perfect hindsight, we might wonder why it took so long for investigators to use nonvocal methods with chimpanzees. Samuel Pepys, in his diary of August 21, 1661, described seeing what he called a "great baboone" on a visit to the London docks. It was probably a chimpanzee, or perhaps a gorilla, although there is no way to be certain after over 340 years. He said, "I do believe that it already understands much English, and I am of the mind it might be taught to speak or make signs." Pepys apparently realized that the sign channel might be superior to the vocal channel, so it is difficult to understand why it was well over 300 years before the Gardners began the first systematic effort to teach signs to chimpanzees. La Mettrie, Robert Yerkes,[16] and probably many others, also suggested the use of signs, but made no serious effort to put the suggestion into practice. Even Charles Darwin[17] acknowledged the power of sign language in 1871 when he said that a practiced signer could "…report to a deaf man every word of a speech rapidly delivered" (p. 21).

David Premack suggested two novel ways to circumvent the vocal channel, and used one of them. The first was a joystick whose manipulation allowed the user to generate the phonemes of English. It was as ingenious, and apparently just as unsuccessful, as Garner's rotating rods.

With apes, three methods have met with considerable success. The Gardners used signs, Rumbaugh used embossed keys (lexigrams) on a computer keyboard, and the Premacks' second method involved magnetized plastic symbols with which their chimpanzees could "write" sentences. Roger Fouts, a student of the Gardners, followed them in using simplified American Sign Language (ASL, or Ameslan), as did Francine Patterson with her gorilla, Koko, and Lyn Miles with her orangutan, Chantek. Thus simplified ASL has been the most popular method with apes.

With marine mammals, Louis Herman and Ronald Schusterman used large-scale arm movements and computer-generated sounds. In every successful case each symbol, whether sign, lexigram, arm movement, computer sound, or plastic form, stood for a word. The only chimpanzee researcher not completely avoiding the vocal channel is Sue Savage-Rumbaugh, who is now encouraging her bonobos to communicate through vocalizations.

Most of the investigators also speak to their animal subjects in English; the exception is the Gardners, who forbade the use of spoken English around their chimpanzees. It is interesting to classify the projects according to whether communication *to* the animal subject was via the same medium as communication *from* the subject. Among those mentioned above, the Gardners stuck strictly to sign. Pepperberg stuck to vocal English. The Premacks and the Rumbaugh-Savage-Rumbaugh projects used plastic symbols and lexigrams, respectively, but English was also used in communicating to their chimpanzees and bonobos. The Foutses and the Patterson/Cohn team used both signs and English. With marine mammals, large-scale signs were used to communicate *to* the animals, and there has been essentially no communication *from* the animals. Thus there has, so far, been no general agreement either about the best method of communicating or about whether other methods must be excluded during training.

WHAT IS LANGUAGE?

There is no consensus about exactly what behavior must be present in order for an animal to have language. There is no standard definition of language that is acceptable to everyone. Both those who think that animals have some language and those who think they have not agree to that. We will propose a definition that implies that animals have learned some language.

At one extreme of the many definitions are very general, abstract definitions that demand very little. For example, we could define language as "Any means of communication between two entities." Such a definition would include all the languages used by computers, such as Basic or Pascal or Lisp, as well as all human and animal methods of communication. This definition might be used by people who are familiar with information theory. According to this theory, information is something that allows one to choose between alternatives. For example, if we must choose between yes and no, and we receive information that allows us to choose correctly, we would say that we had received one bit of information. If the answer were yes, and we then had to choose between now and later, and got the answer to that, we would have received a total of two bits of information.

We can apply this measure to calls of animals, if the calls communicate. Vervet monkeys emit three alarm calls[18,19] in the presence of three different sources of danger (eagles, leopards, and snakes), and monkeys hearing the calls behave as though they too have seen the predators. These three calls provide more than one bit of information, but less than two bits.

Humans can communicate at much higher rates than animals, but our communication with each other via language is much slower than communication between computers. So those who argue that animals lack language because they communicate too slowly might also want to argue that humans lack language because they cannot communicate as rapidly as computers!

We do not suggest that animals or computers are on a par with humans when it comes to language. They are not, and will not be in the foreseeable future, although some computer conversational programs can nearly pass as human conversationalists. However, animal language researchers are not as interested in whether animals can learn language as they are in finding out just what animals can learn about human-designed languages. Using that information, we could then ask how much language each species has acquired. Over the past 40 years or so animals have learned to do more and more complex things related to language, and thus have demonstrated more and more language.

The behaviorist B. F. Skinner[20] proposed a simple definition of language. Skinner used reinforcement as the foundation for everything he believed about behavior, and language was no exception. Language for him was any response that depended for its reinforcement on the intervention of another organism. This ingenious and encompassing definition would include sexual behaviors, as well as requests like "May I have a cookie, please?" Skinner's definition requires no understanding by either the speaker or the listener, and no demonstration that the communication involved syntax or even semantics; all that matters is that the speaker emit some behavior that demands the cooperation of the listener if it is to be reinforced. Reinforcement of a behavior by another organism was, for Skinner, the be-all and end-all of language. Even his severest critic would no doubt agree that Skinner was correct in emphasizing the social, interactional, feature of language.

At the other extreme from Skinner's definition are complex and demanding definitions of language, for example the extensive list of design characteristics of human language proposed by Hockett.[21] For example, human language exhibits "duality of patterning." By this Hockett means that the meaningful units are made up of meaningless units (phonemes, which are somewhat loosely coordinated with letters) that are combined according to certain rules. The meaningful units, in turn, are combined according to other rules into phrases and sentences. Hockett also lists "broadcast transmission" as a property of language, and certainly human speech is broadcast. His design characteristics constitute a very high standard for language, a standard that is difficult or impossible for animals to achieve. Skinner's definition is more "friendly," and it no doubt encouraged researchers to study animals' abilities to communicate.

Bradshaw[22] points out another problem with the criteria for language: they won't hold still! He says:

> Hockett proposed a set of *discriminative* characteristics to distinguish language from animal communication. These characteristics depend not only on our understanding of language, but also on our understanding of animal communication. Since Hockett proposed his criteria, both fields [that is, animal communication and what we think constitutes language] have continued to develop. Several attempts have been made to salvage these criteria through revisions. ... In spite of these revisions, the criteria are once again out of date. ... Revisions of signature characteristics will continue until our understanding of language is complete. The difficulty of duplicating a system that has not been fully described ... hinders our efforts (p. 31).

Bradshaw is therefore concerned that we do not fully understand what language is; certainly there is still disagreement about exactly what the requirements are. As our understanding increases, lists like Hockett's will continue to be revised. There is one additional concern. Language, like organisms, evolves, so it cannot *ever* be fully and finally described.

Having language exhibits some parallels to becoming a person. There are endless arguments about exactly when an embryo achieves personhood, just as there are endless arguments about whether animals have language. In the former case, once a female is pregnant, the embryo is on the road to developing into a person, but when personhood is achieved is open to argument. In the latter case, some animals are definitely well into the embryonic stage in their development of language. The abilities underlying language have unfolded during evolutionary development; so the animals most similar to humans would be expected to have many of the abilities underlying language, and they do. We might say that animals are definitely pregnant with language.

Despite disagreements about the criteria for having language, it is generally agreed that the definition and study of language involve three distinguishable aspects. The first is called semantics, the study of the relationship between symbols and the things they stand for. Loosely speaking, semantics is the study of the meanings of individual symbols. Linguists often talk about representation, which refers to the use of symbols to represent something beyond themselves, or grounding, the process of developing the meanings of words. Thus semantic relationships are absolutely critical for even the simplest language. The evidence that animals can learn the relationship between symbols and their referents is strong.

The second aspect of the study of language, syntactics, is the study of the rules that govern the relationships between symbols. Syntax, roughly speaking, is grammar. An example of a syntactic statement would be that the usual order of words in an English sentence is first, subject, then verb, then object (or other predicate). At present most critics of animal language might agree that animals sometimes understand the semantic relationship between symbols and things, but lack any significant awareness of syntax. Syntax, however, is a step up from semantics, and significant linguistic behaviors can be exhibited in the absence of syntactical rules. Examples of meaningful single-word utterances actually used by animals include things like COKE, CHASE, TICKLE, DRINK, EAT, DIRTY, BIG-TROUBLE, and a host of others. Single-word utterances involve no syntax.

The third aspect of the study of language is called pragmatics. The word refers to the use of language. For example, if a child says "cookie" and receives a cookie, then language has been useful to the child, whether or not the child understands the semantic relationship between the word and the consequences that followed its utterance. There is no doubt that some animals have mastered the use of signs or lexigrams to achieve certain ends. Pragmatics involves the relationship between producers and receivers of symbols, and is thus, like semantics, a fundamental feature of language.

We suggest the following definition of a basic language. We think that the rest of this book shows that what animals have done fulfills the minimal requirements implied by the definition. "Language is an agreed-upon system of signals that represent things, events, feelings, ideas, intentions, and actions on the environment or on other organisms. The signals must symbolize something beyond themselves and fulfill a useful (pragmatic) function by coordinating the activities of organisms. The meanings of the signals comprising a language are shared, at least in part, by the individuals in the group using the language."

Our definition may not be better than others, but we like it and will use it in this book.

WHY STUDY ANIMAL LANGUAGE?

The study of animal language is motivated to a great extent by the ancient challenge to understand the animal mind better.

In 1908, Margaret Floy Washburn[23] said:

To this fundamental difficulty of the dissimilarity between animal minds and ours is added, of course, the obstacle that animals have no language in which to describe their experience to us. Where this unlikeness is greatest, as in the case of invertebrate animals, language would be of little use, since we could not interpret it from our experience; but the higher vertebrates could give us much insight into their minds if they could only speak (pp. 3–4).

Scientists also study animal language because they are simply curious; like other people, they want to know what animals can do. One can imagine an early observer like Richard Lynch Garner being nearly overwhelmed with curiosity and wonder as he tried to teach his young chimpanzee Moses, his gorilla Othello, and his kulu-kamba Aaron, to communicate with him.

A book entitled *The inevitable bond*[24] reveals what happens to most people who do research on animal language—they become as attached to their subjects as most people do to their children, in an extreme version of what happens to most owners of dogs or cats. Early researchers like Garner, however, may have studied animal language because they were seeking reputation and money, as well as knowledge. However, financial motivation cannot explain why so many researchers are so dedicated to the training and testing of their animals.

Most animal language researchers have been couples who dedicated all or part of their lives to the work: examples are the Kelloggs, Hayeses, Gardners, Premacks, Foutses, Rumbaugh and Savage-Rumbaugh, and the Temerlins. These seven couples all worked with great apes, in some cases rearing infant chimpanzees in their homes, and in other cases being available 24 hours a day for the care of bonobos or chimpanzees housed nearby. One could add to them Penny Patterson and Ron Cohn, who have been partners for years in caring for the gorillas Koko, Michael, and Ndume. The exceptions are Nadesha Kohts, who was a single woman when she cared for the chimpanzee, Joni, and Tetsuo Matsuzawa, a Japanese primatologist who has demonstrated single-minded devotion to the 25-year-old Ai, regarded by some as the most intelligent of all the captive chimpanzees.

A third reason for doing animal language research is to find out more about human language, and communication in general. Because scientists need to be extremely careful about the conclusions they reach about animals, researchers working with human use of language have become similarly cautious in their conclusions. Another important practical outcome is that several researchers have used what they learned about animal language to help language-handicapped children. Techniques used with animals may also be helpful with normal children, but the match between animal abilities and the abilities of normal children is not as close as that between animal abilities and those of language-handicapped humans.

Some linguists argue that attempts to teach animals human language reveal little about human language. Whether or not they are correct, there are many fascinating questions that can be answered by finding out how much language animals can learn. These include: What kind of training is most effective with each animal studied, and why? What are the advantages and disadvantages of each of the different methods of communication, for example use of plastic tokens versus sign language? Are methods of training that are effective with animals also useful with language-handicapped humans? What can the study of animal language teach us about the origins of human language? Can special devices that help animals to receive or produce symbols be helpful with humans? Can similarities and differences between human and animal language capabilities provide information about the relationship between brain and cognition? What is the inner world of an animal like?

Attempts to study animal language forced scientists to recognize that they had not fully analyzed the components of language and the way in which humans acquire language. Whether they wanted to or not, they had to break down verbal interactions into their components so that animals could acquire language-like behaviors a step at a time. This provided another question for animal language researchers, that is, "What precisely is involved in human communicative exchanges?" Humans usually acquire language so naturally, without special procedures, that relatively little attention had been paid to the process.

REFERENCES

1. E. S. Savage-Rumbaugh and Lewin, Roger. *Kanzi* (Wiley, New York, 1994).
2. Garner, Richard Lynch, *Gorillas and chimpanzees* (Osgood, McIlvane & Co., London, 1896).
3. N. Kohts, *Infant ape and human child*, 2 vols. Scientific Memoirs of the Museum Darwinianum, Moscow.
4. W. N. Kellogg and L. A. Kellogg, *The ape and the child* (McGraw-Hill, New York, 1993).
5. P. Lieberman, *The biology and evolution of language* (Harvard University Press, Cambridge, MA, 1984).

6. P. Lieberman, *Human language and our reptilian brain* (Harvard University Press, Cambridge, MA, 2000).

7. K. S. MacKain, Speaking without a tongue. *Journal of the National Student Speech Language Hearing Association*, **12**, 46–71 (1984).

8. R. M. Seyfarth and D. L. Cheney, Some general features of vocal development in nonhuman primates: *Social influences on vocal development*, edited by C. T. Snowdon and M. Hausberger (Cambridge University Press, Cambridge, 1997).

9. J. Wind, The evolutionary history of the human speech organs, in: *Studies in language organs*, edited by J. Wind, E. G. Pulleyblank, E. de Grolier, and B. J. Bichakjian (John Benjamins, Amsterdam, 1989), Vol. 1, pp. 173–197.

10. D. K. Candland, *Feral children and clever animals* (Oxford University Press, NY, 1993).

11. R. Allen Gardner, Beatrix T. Gardner, and Thomas E. Van Cantfort, (Eds.). *Teaching sign language to chimpanzees* (State University of New York Press, Albany, NY, 1989).

12. C. Hayes, *The ape in our house* (Harper, NY, 1951).

13. R. Fouts with S. T. Mills, *Next of kin* (William Morrow & Co., NY, 1997).

14. I. Pepperberg, *The Alex studies* (Harvard University Press, Cambridge, MA, 1999).

15. C. A. Ristau and D. Robbins, Language in the great apes: A critical review, in: *Advances in the study of behavior*, edited by J. S. Rosenblatt, R. A. Hinde, C. Beer, and M.-C. Busnel (Academic Press, NY, 1982), pp. 141–255.

16. R. Yerkes, *Almost human* (Century, NY, 1925).

17. M. D. Hauser, *The evolution of communication* (MIT Press, Cambridge, MA, 1996).

18. R. M. Seyfarth, D. L. Cheney, and P. Marler, Vervet monkeys alarm calls: Semantic communication in a free ranging primate. *Animal Behaviour*, **28**, 1070–1094 (1980).

19. T. T. Strusaker, Social structure among vervet monkeys (*Cercopithecus aethiops*). *Behaviour*, **29**, 83–121 (1967).

20. B. F. Skinner, *Verbal behavior* (Appleton-Century-Crofts, NY, 1957).

21. Charles F. Hockett, The origin of speech, in: *Human Communication*. W. H. Freeman, San Francisco (1982; reprint of 1960 article in Scientific American).

22. G. Bradshaw, Beyond animal language, in: *Language and communication: Comparative perspectives*, edited by H. Roitblat, L. M. Herman, and P. E. Nachtigall, (Earlbaum, Hillsdale, NJ, 1993), pp. 25–44.

23. M. F. Washburn, *The animal mind* (MacMillan, NY, 1908).

24. H. Davis and D. Balfour, (Eds.). *The inevitable bond* (Cambridge University Press, Cambridge, 1992).

CHAPTER THREE

Language Research
with Nonhuman Animals:
Methods and Problems

LANGUAGE EVOLVES

If any new anatomic or behavioral characteristic is to evolve, it must provide a reproductive advantage to its possessor. Language cannot be an exception. Language is a method of communicating between individuals, so only species that live socially can benefit from language. Social species can increase their reproductive success by improving their ability to avoid predators; thus species from prairie dogs[1] to monkeys have calls that warn others of the presence of predators. Some animals, notably chimpanzees and bees, direct other members of their social group—especially their relatives—to sources of food. Communication about readiness to mate is also of critical importance. Animals ranging from moths to buffaloes may communicate their willingness to mate chemically through odors, as well as through behavior.

If young animals depend for a longer period on their parents, conditions are better for learning language. Mammals are guaranteed an opportunity to learn communication from their mothers because they will not survive without their mother's milk. A similar situation exists in birds; the young depend on their parents to bring them food until they can fly. For this reason, if for no other, we should expect communication to be more advanced in mammals and birds than in most other groups—and of course that is exactly what we do find.

NATURAL ANIMAL COMMUNICATIONS

One exception to the generalization about mammals and birds is found in bees. They have evolved a complex system for signaling the location of food.[2]

25

Their remarkable food dance evolved in a highly social species whose members are more closely related genetically than are individual family members in, for example, mammalian species. Mammalian brothers and sisters share half of their genes on the average, but worker bees share 3/4 of their genes. Thus in the long run altruistic behavior (communicating about food) benefits worker bees, through survival of their genes, more than it benefits lions or any other mammalian species.

Honeybees communicate the location of distant food sources by a "waggle dance" on the hive. The direction of the dance relative to the top of the hive indicates the direction of the source relative to the sun's position. The length of the run indicates the distance, and the intensity of the dance indicates the richness of the food source. Professor Harold Esch has recently demonstrated[3] that bees gauge the distance to food by keeping track of the number of landmarks they pass during their return to the hive, and demonstrated that the workers who saw the dance perceive the intent of the dancer with remarkable accuracy.

If food is plentiful, bees often do not dance, and if the food source is close, the bees dance in a circle rather than indicating the direction of the food. Although bees also use odor cues to find food, carefully controlled experiments[4] indicate that the dance contains symbolic information. Some people have argued that the bee dance qualifies as a language; whether or not one accepts this argument, bees do have a rather complex communication system that has some features of language.

Knowing how untrained animals communicate provides a foundation for understanding how animals respond to attempts to teach them human language. Elephants have at least 34 distinct vocalizations, many of them in a subsonic range inaudible to humans.[5] Humans are not much better at decoding elephant communications than elephants are at decoding human vocalizations, but the meanings of a few are known; one indicates submission, another that a female is unwilling to mate with a pursuing male. Still others involve dominance relationships.

Animal communications cover several hot topics, including the species, sex, age, and behavioral state of the "speaker." The behavioral state might be a readiness to fight, flee, submit, or mate. Male animals from fish through birds and ungulates to carnivores communicate warnings to other males, indicating a readiness to fight to defend territory and/or the right to mate. Possession of territory provides an increased chance of gaining access to females, with the associated increase in the probability of reproduction.

There are therefore good theoretical reasons to expect the development of communication systems in animals, and observations demonstrating that the reasoning is correct.[6] The nature of the system developed also parallels the conditions of life of the species; for example, it is advantageous for a wide-ranging social organism living with unusually close relatives (bees), all of whom share food, to develop a communication system that helps hive mates find food. However, such a system would not be helpful to solitary wasps, whose food sources are not shared. Similarly, food barks are more helpful to the social chimpanzee than to the less social orangutan.

The way animal communication systems fit a species' conditions of life suggests that human language, too, has an evolutionary basis and precursors in the lineage of human development. The hope for retracing some of this development helped make primates the taxonomic order most in demand for studies of language acquisition, and the chimpanzee in particular, our closest relative, the species to which most attention has been devoted. Knowledge of animals' natural communication systems is also helpful in tutoring animals in sign language; if an untrained animal uses a "natural" sign, that sign can be used as a basis for further training. For example, the Gardners used one of Washoe's natural gestures (extending the arm in the direction of an observer) as the basis for the American Sign Language (ASL) sign for "come." Washoe probably needed to make only a minimal, if any, change in her initial understanding of the meaning of the gesture.

CHOOSING A SPECIES TO STUDY

Making decisions about what animals to use involves practical issues, like the cost of maintenance and the animals' docility, as well as scientific issues like the probable capability of the animal. Jerison[7] and many others believe that the ability to learn is related to brain/body ratio or to the amount of "excess brain" that is not needed to control bodily functions. Very small animals like mice or hummingbirds might have a high brain/body ratio, but they cannot have much excess brain. Chimpanzees and bonobos have brains about a third the size of the human brain. The weight of chimpanzees when well fed is comparable to that of humans (bonobos are smaller, about 85 percent as large as chimpanzees). The brain/body ratios of these species put them in the genius class among nonhuman animals, and they should therefore do relatively well as language students.

Alex, the African gray parrot, belies the expected correlation between excess brain and linguistic ability. His ability to comprehend English and to use a limited vocabulary of English words appropriately is quite remarkable. We must give Alex's trainer, Dr Irene Pepperberg, credit for ignoring the bias against small-brained animals, and overcoming the obstacles involved in working with a species quite unlike our own.

Sea lions (*Zalophus californianus*) and bottlenose dolphins (*Tursiops truncatus*) join chimpanzees as animals that should be highly intelligent. The dolphin brain is approximately as large as the human brain, but most dolphins are larger than humans, so their brain excess is smaller. Also, because they live in the water, as well as lack a human-like vocal apparatus, it is more difficult in some ways to work with marine animals than with primates (although dolphins do not destroy furniture). Nevertheless, investigators like Ronald Schusterman[8] and Louis Herman[9] taught some sea animals to respond to a simple visual receptive language made up of arm and hand signals, and others to respond to a computer-generated acoustic language.

As language learners, the stars of our show are mostly bonobos and chimpanzees. They are also the animals with which we are most familiar. However, other stars include gorillas, orangutans, sea lions, dolphins, and parrots.

CAN ANIMALS LEARN LANGUAGE?

The original big question can be framed as "Can animals learn a human language?" As the eminent linguist Noam Chomsky noted,[10] human language must be distinguished from other communication systems. Chomsky said that the question "Do animals have language?" can never be answered with certainty, because the answer depends upon how we define language. However, he thought that animals clearly did *not* have human language, because the definition of human language was an empirical matter, and we just needed to look to see that no animal does what humans do.

Hockett's[11] thirteen "design features" are often taken as a detailed and specific definition of human language. Several features (e.g., "broadcast communication," "directional reception") are related to the use of the vocal/auditory channel. The features do not always apply well to other channels. Human gestural languages like ASL are now admitted to be full-fledged languages, but are not broadcast like acoustic languages. Another of Hockett's features, total feedback, is also limited to the vocal channel. Signers cannot see the signs they are making from the same perspective as other recipients, and would require a mirror to see some signs.

Partly because of such definitional ambiguities, it is not clear that Chomsky is correct when he says that animals cannot acquire a human language. Certainly animals cannot do what humans do, but the *minimum* requirement for human language is arbitrary. In addition, we do not know what animals will do in the future. Who would have thought 20 years ago that a bonobo, Kanzi, could understand what it meant to have a toy doggie bite a toy snake, with no prior exposure to that sentence?

The realization that a language need not be vocal in order to be a fully human language weakens the case against animal language. The surprising comprehension of English by apes also makes the negative decision less clear. The most competent animals do not absolutely have human language or absolutely lack it. We can, through careful study, learn more precisely what human language-like behaviors they can acquire. The answer to the original question may not be that important anyway. If the question is "Have animals ever learned a human language as we understand the meaning of human language today?" then the answer is "Certainly not." Animals do not, and probably will not, without thousands or millions of years of evolution, write sonnets. Nevertheless, attempts to teach animals human language have opened up an enchanting world, much of which remains to be explored. We have taken the first steps on a spiral staircase of research that began with Richard Lynch Garner, at least in the Western world, and it is still too early to know how high it goes.

There is yet another perspective on the study of animal language. King[12] finds it counterproductive to take a "homocentric" view of animal abilities. Human language should be regarded as a special case of "social information donation," a process that occurs in many species. For example, mothers teach immature animals to mimic their calls, reinforcing correct alarm calls by repeating them, and punishing incorrect calls by physical attacks. King presents several examples of this general type, and believes that accepting animal abilities on their own terms makes it easier to see evolutionary continuities. The question today is "What can animals do as a result of language training?" Or—better because it separates comprehension from production—"What can we communicate to animals through the use of symbols, and what can they communicate to us?"

THE DIFFICULTIES OF ANIMAL LANGUAGE RESEARCH

Researchers in animal language are faced with daunting problems that range from the most mundane to the most philosophical of scientific problems. Let us begin with the host of practical everyday issues. These issues are unavoidable; different animals present different problems, but going from one animal to another merely trades one problem for another.

Consider chimpanzees first, since they are the most-used participants in studies of animal language. Although chimpanzees have often been raised in homes, they make poor houseguests. Catherine Hayes[13] wrote the most amusing description of the tribulations of raising a chimpanzee. She had drapes on her windows when she adopted Viki at a few days of age. As soon as Viki could crawl, Cathy had to cut off the drapes to keep Viki from climbing them. Soon after, Viki was able to jump up and grab the shortened drapes, and they had to be cut again. Soon Cathy had only a valance! In addition, when Viki was only a few months old, she was able to climb up the edges of doors, and not long after was leaping from the top of the kitchen door to the top of the refrigerator.

Chimpanzees, as Viki demonstrated, mature early and have great climbing ability. They are also amazingly strong. A female chimpanzee weighing 135 pounds pulled 1,260 pounds on a dynamometer with both hands, without apparent strain. A 155-pound male pulled 847 pounds one-handed. The average maximum two-handed pull by male college students, several of whom were football players, was 375 pounds. With one hand, the average pull was only 175 pounds. The researchers calculated that the chimpanzees were 4 1/2 times as strong as humans, pound for pound. How they can be this strong, nobody seems to know. The Kelloggs, however, made a comment about their infant chimpanzee, Gua, that suggests a fundamental difference between the muscular compositions of chimpanzees and humans:[14] "By the time the ape has reached the age of 8 months her biceps and other arm muscles *when relaxed* feel nearly as hard to the touch as the *tensed* muscles of many men" (p. 29). Gua was able to chin herself with one hand and climb up and down people by grasping their clothing, with no use of her feet.

It is easy to guess from the above that it is impossible to dominate adult chimpanzees through physical strength, and that they have the potential to be quite destructive. In addition, primates are likely to bite when frustrated, and many people who work with chimpanzees have one or more missing fingers.

Robert Noell, who with his wife raised many chimpanzees, lost two fingers on his right hand and had two fingers on the left severely damaged while working with chimpanzees. Sue Savage-Rumbaugh, the scientist who works most closely with Kanzi, has one short finger, although she is not one finger short (Kanzi did not do it). The famous neurosurgeon, Karl Pribram, lost a finger on a visit to Washoe; she bit it so severely that it had to be amputated. Although she vigorously signed "Sorry, sorry" after the apparently unprovoked bite, the apology was of little help. People at Lemmon's Institute for Primate Studies in Oklahoma lost eleven fingers and one thumb.

Thus, even though several primate laboratories are trying to find homes for a number of adult chimpanzees, you would be well advised not to take advantage of their generosity, even if they would allow you to do so (they would not). Do not bring one home as a surprise for your wife, husband, child, or whomever. A number of chimpanzees have been home-reared in whole or in part of their infancy, but none of them, to our knowledge, has been kept beyond adolescence in a home setting. Luckily for unwanted primates, several private refuges now accept and care for them in semi-natural settings, but unmodified human homes are no place for them.

Gorillas and orangutans present similar problems, although each has a distinctive personality that differs from that of the chimpanzee. In the laboratory, special equipment must be designed that will resist the strength, carelessness and sometimes outright destructiveness of primate subjects. A. Maria Hoyt kept the female gorilla Toto in her home for 9 years. Among the many stories about Toto's strength is the following:[15]

> Once when she seemed annoyed because the station wagon was in her way, she took it by the rear axle with one hand and, though the emergency brake was set, and the entrance to the garage had a slight incline outwards, yanked the car out and with a quick thrust of her arm dashed it forward again into the garage wall, smashing the headlights (p. 178).

The point is clear: Do not frustrate a 9-year-old gorilla, even a female (males are larger and stronger). On another occasion, Toto, in an excess of enthusiasm, knocked Mrs Hoyt down and broke both her wrists. Toto repeatedly bit her favorite caretaker, Tomas, who was forced to wear heavy leather gloves and arm sheaths in her presence. Mrs Hoyt reports that "I have said—and surely in a large sense it is true—that a gorilla is incapable of self-discipline (pp. 185–186). It is certainly understandable that, despite her love for Toto, Mrs Hoyt placed her "baby" (who weighed 438 pounds) with the circus, where she became a companion of the famous male gorilla, Gargantua.

Orangutans, too, are forbiddingly strong and difficult to handle as adults, despite their placid and slow-moving appearance. When pressed, they can move quickly, and are amazingly strong. In an experiment at the San Diego Zoo, an adult male easily lifted with one arm a weight that a trained weight lifter could not raise with both arms and his feet braced against a wall. Thus researchers cannot demand the cooperation of any adult ape; the ape must truly volunteer her or his services to the experimenter!

A problem that is associated with the size and strength of primates is the cost of feeding and housing. Herbert Terrace became so discouraged by the continuing struggle to obtain funding that he sent his chimpanzee subject, Nim, back to Oklahoma at an early age. The Gardners and the Foutses had similar problems. Dolphins and sea lions are also expensive to maintain. In this respect, parrots have much to offer!

INTERPRETING ANIMAL LANGUAGE: GLOSSING AND GROUNDING

We turn now from the practical problems associated with working with animals to the more technical problems of interpreting the results of observations and experiments on animal language. We immediately encounter the problem of meaning; that is, what does an animal really mean by pressing a key with a lexigram that we take to mean *give*, or making the correct sign for give according to ASL, or selecting a piece of plastic that we intend to mean give, or making an instinctive begging gesture that we interpret as give, or even saying give in English? When we say that we take any of these manifestations as give, we are glossing the manifestation as though it were equivalent to give as we understand it. Would it be as accurate to gloss the symbol, as in "I would like to have that, please?" If your goldfish, like mine, swim over to your side of the pool each evening when you show up, should their behavior be glossed as "Please, owner, give fish food?" You would, of course, think that pretty silly; the swimming is more simply interpreted as a response elicited by the circumstances surrounding the presentation of food and, so far as we know, the behavior has no communicative intent. When Sue Savage-Rumbaugh titled her book *Ape language: From conditioned response to symbol*, she probably had this kind of distinction in mind. So presumably did the Gardners when they subtitled their book *From sign stimuli to sign language*. The investigator of animal language must be very careful to find out whether the animal is communicating (in which case the behavior can be taken as symbolic) or merely performing a learned behavior (in which case we speak only of conditioned or instrumental responses).

Investigators of animal language cannot claim that animals exhibit linguistic ability unless the claim is necessarily implied by their observed behaviors. This obligation follows from Occam's Razor, a very old and revered philosophical principle attributed to William of Occam (ca. 1280–1349), although it was in fact used by many still earlier thinkers. Occam's principle is also called the principle of parsimony, and it stipulates that no more things should be presumed to exist than are absolutely necessary. C. Lloyd Morgan applied Occam's Razor to psychology by stating that no behavior should be interpreted as a consequence of a function higher on the psychological scale if a function lower on the scale was sufficient to explain it. Morgan was apparently not troubled by the need to provide an ordered scale of psychological functions, which looked suspiciously like a scale with exclusively human functions at the top, and other functions arrayed below.

As applied to the study of language, Morgan's statement forbids the presumption that animals understand or are producing language if some simpler, more fundamental, process can account for the behavior. We already applied this principle in the previous paragraph when we said that we could not assume that linguistic capability was implied by a behavior if the behavior could also be accounted for by conditioning. Thus Dr. Savage-Rumbaugh's and the Gardners' titles not only specifically recognize the problems peculiar to demonstrating language in chimpanzees, but also constitute obeisance to Lloyd Morgan's canon and Occam's principle of parsimony.

As an example of the application of Lloyd Morgan's canon, let us take an observation of the behavior of Rumbaugh's chimpanzee, Lana. Lana became very adept at pressing a series of lexigrams in order to obtain a food reward.[16] A sample series of six lexigram keys might be glossed as PLEASE MACHINE GIVE PIECE-OF APPLE PERIOD. The question is whether Lana is writing a proper sentence; the series is rewarded with apple only if the keys are pressed in the appropriate sequence, so it is grammatical in that sense. However, would Lana do just as well if we glossed the keys as FIRST SECOND THIRD FOURTH FIFTH SIXTH, connoting that Lana has just learned to press a sequence of six keys that has been rewarded in the past? Lana need know nothing of the meaning of the word "please," for example, although the key representing that word for humans had to be pressed in order to obtain her reward. And it is highly unlikely that she has a conception of the word "machine" that is at all human-like. So the most parsimonious explanation of her behavior, if we know no more than is stated above about her capability, is that she has learned a chain of operant responses in order to obtain a reward. The words that we use as glosses for the lexigrams need not mean anything to Lana. It should be noted that Rumbaugh made no claim that the PLEASE or PERIOD keys had any meaning for Lana beyond initiating and stopping the sequence; he claimed only that the other keys had semantic significance. Lana showed semantic knowledge by using the same keys in different orders and different combinations, for example, by substituting Tim for machine in the sentence above and pressing lexigrams for PLEASE TIM GIVE LANA APPLE PERIOD.

If Lana could use the lexigram keys to name objects presented to her—for example, machines or apples—she would be exhibiting behavior that could not be accounted for in terms of operant chains; if she could classify several related lexigrams into classes, for example by labeling them all as foods, her behavior would look still more like language. Thus as the complexity of the behavior increases, the complexity of the functions that underlie it may also have to increase. Lana did learn to name, rather than request, objects denoted by the keys, although she required additional training to make the step from requesting to naming. Several other chimpanzees and bonobos have subsequently displayed naming and classifying behaviors demonstrating that various tokens of language—ASL signs, plastic shapes, and lexigrams—function as symbols rather than simply as conditioned stimuli.

Children's responses to questions sometimes indicate that the meanings of words for them lie between the word's meaning for an animal and its meaning for an adult human. For example, you could ask a child, "Why do they call December 25 Christmas day?" and the child might answer, "Because you get lots of presents." It is clear from this response that the child's meaning differs from that of most adults in our culture. If animals could answer questions about the meanings of their communications, the answers might surprise us. Even if an animal attaches meaning to the word Christmas (and Koko and the Foutses' chimpanzees, and perhaps other animals, celebrate Christmas) he or she would not have as rich a set of associations to Christmas as a 3-year-old human child, and therefore Christmas almost certainly means less to an educated chimpanzee or gorilla than to a child.

Premack dedicates much of his book, *Gavagai*,[17] to the question of under-standing what an animal (or a human, for that matter) means by a behavior that may or may not be linguistic. The title of his book refers to yet another book, in which the philosopher Willard Quine asks what would be meant if a native of some other country, speaking in an unknown language, exclaimed "Gavagai!" when a white rabbit jumped up from a hiding place and ran. Quine and Premack point out that we could not tell what the word meant. Our first thought would be to gloss *gavagai* as *rabbit*. But it might mean *white* or *food* or "it's running" or "look at that," or "shoot him with the bow and arrow you're holding in your hand," or even "Damn, that startled me!" We require many repeated exposures to *gavagai*, or to any other word, in different contexts before we can ascertain its meaning. If someone says "gavagai" again upon seeing a brown rabbit in a cage, we suspect that the word really does mean rabbit. If, however, he says it the second time when he sees a clean linen sheet, we suspect that he meant *white*. If the second time occurs in the presence of a mango, we suspect that it means *food* (or *delicious*, or *disgusting*). If we hear the word for the second time when a bull African elephant charges trumpeting out of the bush, we quickly reject the rabbit, white, and definitely the "shoot him with the bow and arrow" ideas, and come down on the side of being startled (to put it mildly)!

There are, therefore, at least two aspects of glossing. The first is the question of whether a potential symbol, regardless of how it is expressed, should be taken as a symbol at all. The second is, given that the possible symbol really is a symbol, what does it mean to its user?

We are so used to our native language that our first inclination might be to rush to the dictionary to look up the meaning of whatever symbol is under consideration. However, for animal language studies, we would immediately be stymied; very few of us would understand a dictionary that explained the signs of ASL by using other signs, and there are no dictionaries that explain the arbitrary lexigrams used by Rumbaugh by using other lexigrams, or the shapes of the plastic words used by David and Ann Premack by using other plastic shapes.

This brings us face to face with the other side of the glossing problem. This new side is called the grounding problem. We have no grounding in lexigrams or plastic symbols that will let us use simpler symbols to define more complex ones. Precisely the same problem is encountered in our native languages, and is solved by us as we learn what words mean by observing how those around us use them. Once simple words are grounded, we can use an English dictionary that defines unknown new words by using these simple words. I can, for example, look up D ration and find that it is "a United States Army emergency field ration consisting of a specially prepared chocolate bar having a highly concentrated food value." I then think I understand what D ration means because I have grounding in chocolate, and U.S. Army and so on; however, I still would not recognize a D ration if I saw one, nor should I delude myself by believing that the meaning of D ration has now become the same for me as it is for a person who subsisted on these specially prepared bars during the Battle of the Bulge near the end of World War II. I do not have, nor do I wish I had, that kind of grounding for the meaning of D ration. I neither know how they taste, nor what "specially prepared" really means, nor how well the highly concentrated food value satisfies hunger when eaten in a foxhole after a week under continuous enemy fire.

Although this example is a special case, so are nearly all examples of unknown words. It is unlikely that any word means exactly the same thing to two different people. Different grounding creates different meanings. The same logic applies to word combinations; for example, Dr. Hillix's wife, raised in the city, did not know the literal meaning of "looking a gift horse in the mouth." Raised on a farm, Hillix knew that one could tell a horse's approximate age, and therefore value, by looking at its teeth. Although meanings may differ, without grounding words have *no* concrete meaning; they are not symbols unless they symbolize something. Thus it must be that people communicate more accurately with each other if they have had experiences that grounded the meanings of words similarly. The same logic applies to both people and nonhuman animals. If they have had similar grounding with a particular symbol, we can reasonably expect that there may be some core of meaning that is shared by human and animal.

However, there are severe limits on that reasonable expectation. One problem is that humans have had extensive experiences with language itself; the word *give* (or other symbol glossed as *give*) has been read as well as heard many times, and has taken its place in an extensive linguistic framework. We know that you can give HIV, as well as an ice cream cone. Most Americans know that the infinitive form is to give, that the past tense is gave, that the third person singular present tense form is gives, and so on. Although many people are not consciously aware of these grammatical niceties, they nevertheless tend to use the various forms of give correctly, and thus to demonstrate that the word has a place in their grammatical network. Nonhuman animals have, at most, a much less extensive network; Kanzi, the most accomplished nonhuman animal, appears to have a network comprising several hundred words. Yet Kanzi's grounding of individual words in concrete events differs radically from that of the average human listener. His grounding in grammatical relationships no doubt differs far more radically.

Sue Savage-Rumbaugh tells an amusing story about Kanzi's grounding, in more ways than one, of the meaning of an electrical outlet.[18] Sue had vehemently explained to Kanzi that electrical outlets were dangerous, but that only piqued his curiosity. Like any self-respecting primate male, he took it as a challenge. When Sue was occupied elsewhere, Kanzi stole a screwdriver and stuck it into the outlet, with the expected result. "He stood ramrod straight, and his hair rose two inches. He yanked the screwdriver out of the socket and immediately burst into a series of emphatic 'Waa' sounds" (p. 8). Kanzi then picked up anything he could get his hands on—balls, blankets, and toys among them—and threw them at the outlet. After this grounding experience, a lexigram for ELECTRICAL OUTLET, as we would gloss it, would have a very different meaning for Kanzi than for a human who understood electricity and its many uses (beyond delivering shocks).

Even if human and nonhuman animals share what appear to be identical groundings for a symbol, clear problems remain. Different organisms experience identical contexts differently. Imagine, for example, how different life must be if one is wearing a collar attached to a leash, versus holding the other end of the leash, let alone if one's brain and sensory systems are different from other observers' brains and senses, as is very much the case with humans versus dolphins.

If a non-vocal language is used in teaching the nonhuman animal, the nature of the grounding problem undergoes some change. Consider, for example, teaching Washoe ASL. The Gardners and most of their students did not know ASL when they started to teach it to Washoe, and the use of vocal language around Washoe was kept to an absolute minimum. Only when, for example, a refrigerator repairman came was vocal language used. Thus Washoe's grounding was directly in ASL, and her teachers were receiving similar grounding at the same time. English glosses were used only for purposes of discussion among the researchers (outside of Washoe's presence) and for clarity in published descriptions of the work. Under these conditions the meanings of the ASL gestures should have been more similar for human and ape than is the case when vocal language is used in animal language research.

Similar arguments can be made for Rumbaugh's use of lexigrams on a computer keyboard and Premack's use of plastic symbols. The human researchers and chimpanzee participants received similar grounding with these unique symbols. However, it is likely that the human participants glossed—translated—the symbols into their English equivalents, thereby embedding the glosses into their grammatical networks, with their complexity of interrelationships. There is probably no way to avoid this problem completely, although the Gardners showed good awareness of it and did what they could to avoid it by refusing to allow the use of English in the presence of the chimpanzees.[19]

The old problem that there may be no exact equivalence between words in different languages occurs because the experiences connected with a particular word in one culture never correspond precisely with the experiences connected with *any* word in a different language. An example given to me was the word *gezellig* in Dutch. It might be translated as cozy, or warm, or intimate, but people who know both languages very well say that there is no word in English that expresses exactly what a Dutch speaker means by *gezellig*.

When glossing the symbols used in animal language studies, the best we can hope for is to find or create a dictionary that translates ASL, or Rumbaugh's Yerkish, or the Premacks' plastic symbols, or the signs used by Herman with dolphins or Schusterman with seals, into English (or some other vocal language). But even with such a dictionary, we encounter the usual problems of inexact correspondence, plus—in many cases—the problem of crossing from a visually to a vocally based language.

Let us now return to the grounding problem, armed with awareness of how dangerous it is to assume that an animal's use of a symbol is the same as our use of an English gloss of that symbol. Chomsky, as you might expect, was very concerned with this possible error. He pointed out that Straub, Seidenberg, Bever, and Terrace were able to teach pigeons to peck a sequence of four keys in order to obtain food; that is similar to what Rumbaugh taught Lana to do on her lexigram board. Chomsky[10] then asks:

Suppose that we label these buttons, successively, PLEASE, GIVE, ME, FOOD. Do we now want to say that pigeons have been shown to have the capacity for language, in a rudimentary way? The Gardners, in an article reviewing such work, argue that virtually all of it apart from their own is undermined by a false analogy, in that researchers have labeled the symbols taught to apes with values derived from human languages, as in the pigeon example, then mistakenly concluding that the symbol correlated by the human researcher with a term of human language is being used by the ape with the properties of its human language correlate (pp. 434, 437).

Chomsky does not consider the many additional steps taken by ape language researchers to demonstrate that the apes' symbols are used referentially. He goes on to argue that the Gardners are as vulnerable to their own criticism as are the investigators they criticize. "The question arises in exactly the same form under the Gardners' approach (in which the similarity is based on visual similarity or even identity) as in the work of Premack, Rumbaugh, etc., which they criticize" (p. 437).

COMPUTER UNDERSTANDING

The question posed by Chomsky and the Gardners is clearly important. A similar problem is raised in a different context, that of computer understanding, by J. R. Searle in his Chinese room thought experiment.[20] Searle argued that computers could seem to understand language although they understood nothing, because of the way the computer's responses were interpreted by observers. Searle's argument could also apply to animals.

The thought experiment goes as follows. Imagine that you are inside a room with a small slot into which people can pass questions written in Chinese characters. These characters are completely meaningless squiggles to you. However, you have a full set of English instructions that tells you how to associate the squiggles passed in with the correct squiggles to pass out. You understand no Chinese at all, but observers who pass the Chinese characters in and read the ones you return will assume that you know Chinese.

This is similar to the argument presented with respect to animals by Chomsky and the Gardners. In both cases the entity inside the room does not understand the symbols in the same way as the entities outside the room. Does that mean that, no matter what an animal or computer does, we cannot assume that it understands language?

I do not think so. If a computer, a person inside a Chinese room, or an animal were not guided by some system analogous to language, it would be unable to fake the responses. The system in which the entity worked would have to be as complex, and in some sense as linguistic, as the system in which it appeared, from the outside, to work. If the entity inside had learned its language in the same way as the entities outside the room, the grounding of the insider's symbols would be similar to that of the outside observers. Surely grounding that was completely identical with that of the external observers on the part of animals, or computers, or the person inside the room, would be impossible to demonstrate. However, it would be equally difficult to demonstrate identical grounding between any two of the outside observers. Searle is correct in saying that the computer, or the person inside the room, need not understand Chinese. But we know that the computer or the person at least in some sense understands the language in which they were instructed in how to associate symbols! In the same way, we would have to presume that an animal that carried out all the behaviors expected of a language-competent entity understood some set of instructions adequate to producing the behaviors.

Dennett[21] reached a conclusion similar to ours after analyzing whether Searle's entity in the room could really carry out the task Searle posed for it, sans any grounding or understanding of the language. Dennett argues that it is quite impossible to make sense of strings of symbols without having not just grounding but also an understanding of analogies, metaphorical usages, irony and all the rest that constitutes our human understanding of language. This line of argument presumes that no animal or computer has the kind of understanding that Dennett demands, so Searle's point might still be well taken with respect to animals' understanding of language. However, the bonobo Panbanisha, a stepsister of Kanzi, once called a woman she liked mushroom, because she had a hairdo that resembled a mushroom. Critics may call this a coincidence, but Panbanisha's metaphorical intention seemed to be very clear, and thus to demonstrate the kind of capability that Dennett said was required. She had no training to label a human with a lexigram that meant, literally, mushroom!

The Gardners had a reasonable basis for their claim that there was more than the usual degree of agreement between their interpretation of the meanings of signs and their animals' probable interpretations. They taught sign language directly to their chimpanzees. Washoe was taught her signs directly, somewhat as a learning-handicapped deaf child might be taught. The researchers may have been handicapped by their tendency to gloss signs as having ASL or English meanings—but that should not detract greatly from our interpretation of what Washoe understood, for Washoe was grounding signs in her own experiences, experiences that the researchers, for the most part, shared. Chomsky and the Gardners knew that the signs used by the Gardners were "correlated with ASL," and with English, but Washoe knew nothing of that correlation. Chomsky was correct in that it is easy for experimenters, and for readers of articles in which English meanings are given for the various kinds of tokens used, to assume that the meanings of the tokens as used by animals are the same as the meanings of the glosses. However, to conclude from that that the animal has manifested no understanding of language is wrong.

It is ironic that the very feature of the Gardners' research for which they were severely criticized was an advantage from the point of view of shared grounding—that is, very few of the people who worked with Washoe were fluent signers during the early stages of the research.

CONCLUSION: SEMANTICITY IS CRITICAL

The conclusion must be that understanding a language requires that the speaker have semantic references for the symbols—that the symbols be grounded in the real world. At the same time, we should recognize that the entity in the room, whether the room is a Chinese room, a computer case, or animal (including human) skin, must know *something* if it interprets or uses signs in a way that, from the outside, looks like language comprehension. The English-speaking inhabitant of Searle's Chinese room knew English, and the English *language* conveyed the information about the squiggles. A computer or an animal that could perform the same squiggle-matching task would have to know computer, or chimpanzee, or have mastered some other complex procedure, in order to give the appearance of knowing Chinese.

Those of us who speak, write, and appear to understand English are doing so via some physiologically based legerdemain which may be every bit as mysterious as the technique of the English speaker would be to Chinese observers outside the room who knew nothing of English. Chomsky thinks that we speak and understand our native languages through reversible transformations between deep and surface grammatical structures. Perhaps; but all we know for sure is that a user of language behaves in the world as though he or she understands the language. It may be best to recognize this, and to worry less about whether the language means the same thing to animals as to humans. The assumption that two entities mean the same thing is always problematic, even when the external observer and the entity are two native speakers of the same language.

It therefore appears that we have as much right to accept the apparent understanding of some language by some animals as to accept similar understanding in human children or handicapped adults. Computers, by virtue of their different grounding, understand only the syntactic relationships between symbols. And the symbols are not symbolic *of* anything outside the set of symbols for computers, so *the symbols are not symbols* if they do not refer to anything outside the symbol set.

CHANGING PERSPECTIVES ON ANIMALS

Language experiments with nonhuman animals have contributed to a radically different perspective on our human relationship to, and responsibility for, animals. Animals are capable of much more, intellectually and emotionally speaking, than most people believed 50 years ago. The depth of intellect and emotion that they have demonstrated has increased our respect and empathy for animals.

Despite her love for Toto, Maria Hoyt, 70 or so years ago, expressed a very different attitude when she described her husband's accomplishments with pride:[22]

> ... there are few white hunters who have bettered Kenneth's bag of twenty-three African, three Indian, and three Indo-Chinese elephants. As recently as 1934 ... his African fever had not subsided. He received a letter from Mr. Reel, the white hunter who had been with us when we acquired Toto, saying that one of the largest elephants ever sighted had been seen in the Ivory Coast region ... he came back with the tusks of the great creature he had sought, carefully hewn from the skull by the picturesque Pembe Moto, and each weighing more than a hundred pounds (pp. 171–172).

What gave Mrs Hoyt great pride in her husband is now an abomination that makes you shudder as you read about it, just as the abuse of animals in medical studies, language studies or any other venue is an abomination. Killing elephants, chimpanzees, gorillas, or other "bush meat" as an act of personal desperation, brought on perhaps by impending starvation, is now in many places poaching that might cost a greedy or hungry poacher his life. Murdering an elephant to bring home its tusks for a trophy is now both illegal and unthinkable.

The men and women who study animal language all take pride in anything they can do to preserve species, not kill them. Animal language research has made a rich contribution to the new awareness that animals share more of our intellect and feelings than we previously believed, and therefore deserve protection. Thus this research deserves support because it contributes to the richness of animal and human life.

REFERENCES

1. J. Kiriazis and C. Slobodchikoff, Anthropomorphism and the study of animal language, in: *Anthropomorphism, anecdotes, and animals*, edited by R. W. Mitchell, N. S. Thompson, and H. L. Miles (SUNY Press, Albany, NY, 1997), pp.365–369.

2. K. von. Frisch, *The dance language and orientation of bees*. (Translation by L. Chadwick.) (Harvard University Press, Cambridge, MA, 1967.)

3. C. Rist, Honeybee dance. *Discover*, **23**, 1, 81 (2002).

4. C. D. Michener, *The social behavior of the bees* (Harvard University Press, Cambridge, MA, 1974).

5. D. H. Chadwick, *The fate of the elephant* (Sierra Club Books, San Francisco, 1992).

6. M. D. Hauser, *The evolution of communication* (MIT Press, Cambridge, MA, 1996).

7. Harry J. Jerison, Animal intelligence as encephalization. *Philosophical Transactions of the Royal Society* (London), **B308**, 21–35 (1985).

8. R. J. Schusterman, J. A. Thomas, and F. G. Wood, (Eds.). *Dolphin cognition and behavior: A comparative approach* (Erlbaum, Hillsdale, NJ, 1986).

9. L. M. Herman, D. G. Richards, and J. P. Wolz, Comprehension of sentences by bottlenosed dolphins. *Cognition*, **16**, 129–219 (1984).

10. N. Chomsky, Human language and other semiotic systems, in: *Speaking of apes: A critical anthology of two-way communication with man*, edited by T. A. Sebeok and J. Umiker-Sebeok (Plenum, NY, 1980), pp. 429–440.

11. Charles F. Hockett, The origin of speech, in: *Human Communication*. W. H. Freeman, San Francisco (1982; reprint of 1960 article in *Scientific American*).

12. B. J. King, Evolutionism, essentialism, and an evolutionary perspective on language: Moving beyond a human standard. *Language and Communication*, **14**, 1, 1–13 (1994).

13. C. Hayes, *The ape in our house* (Harper, NY, 1951).

14. W. N. Kellogg and L. A. Kellogg, *The ape and the child* (McGraw-Hill, NY, 1933).

15. A. Maria Hoyt, *Toto and I: A gorilla in the family* (Lippincott, Philadelphia, 1941).

16. Duane M Rumbaugh, (Ed.). *Language learning by a chimpanzee: The Lana project* (Academic Press, NY, 1977).

17. D. Premack, *Gavagai! or the future history of the animal language controversy* (Bradford, Cambridge, MS, 1986).

18. S. Savage-Rumbaugh, S. G. Shanker, and T. J. Taylor *Apes, language, and the human mind* (Oxford University Press, NY, 1998).
19. R. A. Gardner and B. T. Gardner, A cross-fostering laboratory, in: *Teaching sign language to chimpanzees*, edited by R. A. Gardner, B. T. Gardner, and T. E. Van Cantfort (State University of New York Press, Albany, NY, 1989), pp. 1–28.
20. J. R. Searle, Minds, brains, and programs. *Behavioral and brain sciences*, **3**, 417–424 (1980).
21. D. Dennett, *Consciousness explained* (Little, Brown, Boston, 1991).

CHAPTER FOUR

Early Reports about Language in Animals

EARLY MYTHOLOGY

Animal language has been a subject of discussion in myth and fables for thousands of years. In about 1,000 BC the author or authors of the *Book of Genesis* reported a conversation between Eve and the serpent, which allegedly accounts for humans' ejection from Paradise. As reported in 500 AD by the scribe Horapollo Nilous,[1] Egyptian priests from ancient times believed that some baboons had the power of reading and writing, and therefore gave baboons newly arrived in the temple a test of their language abilities. They gave each baboon a quill pen, ink, and a tablet. If the baboon was deemed capable of writing, it was regarded as sacred, fed wine and choice roast meats and not required to work. Other baboons got menial tasks to perform and no special diet. Unlikely as Nilous's account seems, it is supported by the discovery of mummified baboons, most with rickets that a diet of wine and roast meat might cause! Thus Nilous reported the first test of animal language ability 1,500 years ago, and the test may have been hundreds of years old when he wrote about it.

These ancient accounts of talking serpents and sacred baboons reach far beyond the bounds of modern credulity, but they serve as examples—though much exaggerated—of humanity's apparent need to seek nonhuman company, whether of gods, extraterrestrial visitors or animals. Animal language researchers, with a great deal of help from their eager critics, remain on guard against the anthropomorphic tendencies revealed in these early myths.

Animals that talked were not limited to the religious literature, of course. Beginning around 600 BC, Aesop wrote his fables about talking animals, prominent among whom was the famous fox whose inability to reach the grapes gave us the model for rationalization ("the grapes were probably sour anyway"). That tradition reaches from Aesop to the present, in which the Dr. Dolittle stories of Hugh Lofting and the children's books of Dr. Seuss continue the ancient tradition.

CREDULOUS AND INCREDULOUS PHILOSOPHERS

Philosophers, like novelists, have long toyed with the idea of talking animals. John Locke, a famous English philosopher of the late 16th and early 17th centuries, worried about the status of animals with human intelligence, or humans without it. He says that a being shaped like a man would be considered a man still, even if very unintelligent, and an intelligent cat or parrot would be considered just that, not a man. Locke continues:[2]

> A relation we have in an author of great note, is sufficient to countenance the supposition of a rational parrot. His words are: "I had a mind to know, from Prince Maurice's own mouth, the account of a common, but much credited story, that I had heard so often from many other, of an old parrot he had in Brizil [sic], during his government there, that spoke, and asked, and answered common questions, like a reasonable creature: … he told me … that he had heard of such an old parrot [and] that he had so much curiosity as to send for it … and when it came first into the room where the prince was, with a great many Dutchmen about him, it said presently, *What a company of white men are here!* They asked it, what it thought that man was, pointing to the prince. It answered, *Some General or other.* The parrot was asked, "Whence come ye?" It replied "From Marinnan." The Prince asked "To whom do you belong?" The parrot replied "To a Portuguese." "What do you there?" asked the Prince. The parrot answered "I look after the chickens." The Prince laughed, and said "*You* look after the chickens?" The parrot answered "Yes, I; and I know well enough how to do it" (p. 436).

This story, of course, stretches credulity, and we feel safe in asserting that the story is very unlikely to be true in its entirety. However, the African gray parrot Alex has demonstrated amazing abilities that make us more prone to believe that a parrot might at some earlier time have shown considerable intelligence. It is difficult for scientists as well as for laymen to maintain openness to new phenomena without sometimes being taken in by charlatans. We will never know how much if any of the third- (or higher) hand story told by John Locke was based on truth. Because science demands proof, we must take the skeptical path and regard the tale as unproven, while recognizing that we run some risk of being wrong.

Richard Lynch Garner

It was 306 years after Pepys' suggestion before attempts to teach animals sign language emerged as a serious enterprise. It was about 230 years before the first systematic attempts were made to teach chimpanzees any form of language. That was in the latter part of the 19th century, at about the time that psychology was emerging as a science.

Richard Lynch Garner in 1892 published a book[3] on the language of monkeys, and went to Gabon the same year. There, 2° N of the equator, he built a cage in which he stayed for protection against wild animals. Not long after that, two men, Herr von Osten and Karl Krall, began to work with horses.[4] All three men eventually provided their animal subjects with some way to spell out words, although so far as we can tell none of the animals involved had the slightest ability to spell unless they were cued by humans. Thus, in avoiding one problem (bypassing the vocal channel) they all fell into another (asking animals to construct words).

Expecting animals to spell seems incredibly naive to any parent who has ever resorted to spelling in order to keep a secret from a child. My mother loved to tell the story of when I, during my second year, first broke the code. My father asked what was for dessert, and my mother, knowing my proclivity for eating nothing else when pie was in the offing, responded "p, i, e." I smiled and said "t, i, e, pie!" When the bonobo Kanzi began to comprehend English, Sue Savage-Rumbaugh had to revert to the parents' trick:[5] "... we had to do what many parents do when they don't want their children to overhear; we began to spell out some words around Kanzi. Kanzi, like most children, recognized that we were doing this to avoid his listening and simply began to listen all the harder" (p. 149).

Garner's 1896 book[6] on his experiences with gorillas and chimpanzees is a wonderful mixture of observation and myth, and is thus a great transition from early mythology to modern science. Garner does his best to eliminate the mythology of gorillas as terrible and dangerous animals. He does say that they eat meat—his young captive gorilla Othello liked meat, but Garner also claims that gorillas eat rodents, lizards, toads, and even porcupines. He correctly says that chimpanzees like meat better. Garner incorrectly says that neither species builds nests; no wonder he thinks so, since his observations were gathered largely while in his cage, and neither species built any nests in his cage!

The natives told him that gorillas built shelters, but Garner quite properly refused to believe them because he could find no direct evidence in his travels (even outside the cage). Garner thus gives us six kinds of evidence: correctly interpreted direct observations, incorrectly interpreted direct observations, correctly accepted reports, incorrectly accepted reports (for example, that gorillas eat porcupines), correctly rejected reports, and incorrectly rejected reports (for example, refusing to believe that gorillas or chimpanzees build nests). Garner's work is, therefore, valuable pioneering work, but his conclusions must be regarded with considerable skepticism.

Garner believed that chimpanzees had words for *food, good, danger* or *strange* and *come*, which he and they understood and shared. These were native chimp sounds. He also thought he understood several other sounds whose nature he was not able to describe adequately, and yet others whose meaning he could not decipher. Garner also describes natural chimpanzee gestures, including the one that the Gardners later used for *come*, with Washoe; Garner aptly says that it consists merely of extending the arm, without motion of the wrist, toward the person or thing desired.

Garner's favorite chimpanzee was Moses, although he was even fonder of his kulu-kamba, Aaron (Modern writers spell it *koolakamba*). Moses loved to play peek-a-boo with Garner and his native boy—many chimpanzees enjoy this game. Moses also loved canned beef. He learned the purposes of tools like can openers, but was not good at using them. He tried unsuccessfully to saw with the back side of a saw, but could and would use a file on anything he could reach.

In trying to teach Moses to speak, Garner took him on his lap and bribed him with the favored canned beef. He thinks Moses understood what was wanted from him; he tried to imitate *mamma* but merely imitated the lip movements without the sound.

Garner also tried to teach his gorilla, Othello, to speak. Othello had a different version of the voicing problem, the same one later encountered by the Hayeses with Viki; Othello tried to make noises on inspiration of breath, rather than on expiration. Garner trained Moses for less than 3 months, ending when he was a little over a year old. Moses died at this time. We now know that it was naive to expect Moses to learn any speech in that short period.

Garner left Moses in the care of a missionary who got a fever and, in turn, left Moses in the care of a native boy. Moses then contracted a respiratory disease that Garner could not cure.

Finally, Garner tried to teach his kulu-kamba Aaron to speak, and thought he got a fair rendition of the French word for fire, *feu*. He tried to get *mamma*, but had little success. Garner developed a notation with which he believed he could express all the sounds made by humans and other animals; it expressed the positions of the parts of the vocal apparatus, with ( ), for example, representing the wide-open position of the glottis, which should produce the sound of *O*. The symbol (.) was to represent a half-closed glottis, (German *U*), and (:) a closed glottis to produce an *A*. Garner's notation was primitive, and he was apparently unaware of the extent of the differences between animal vocal tracts. He was, however, not completely oblivious to differences; he stated that chimpanzees have two air sacs on the insides of their throats, which act as reservoirs and amplifiers for sound, making it possible for chimpanzees to be noisier than people.

His final claim was:[6]

In conclusion, I will say that the sounds uttered by these apes have all the characteristics of true speech. The speaker is conscious of the meaning of the sound used, and uses it with the definite purpose of conveying an idea to the one addressed; the sound is always addressed to some definite one, and the speaker usually looks at the addressed; he regulates the pitch and volume of the voice to suit the condition under which it is used; he knows the value of sound as a medium of thought. These and many other facts show that they are truly speech.

It is remarkable that Garner feels free to speak thus, but one must remember that he was writing many years before the advent of a more skeptical scientific viewpoint. In Garner's time subjective judgments were more acceptable than they are now, more than 100 years later.

Garner got his kulu-kamba, Aaron, on the trip he was taking when Moses took sick. He attributes greater intelligence and sympathy to Aaron; later in his book he devotes much time to speculation on what a kulu-kamba really is, maybe an offspring of a gorilla father and chimpanzee mother, or vice versa, both according to different versions by natives; or maybe a unique species, or just superior individuals of the chimpanzee species, but in any case more intelligent than most chimpanzees.

Whatever kulu-kambas are, Garner certainly had a high opinion of them:[6] "If proper conditions were afforded to keep a pair of kulus in training for some years, it is difficult to say what they might not be taught. They are not only apt in learning what they are taught, but they are well-disposed, and can apply their accomplishments to some useful end" (p. 187).

The confusion about kulu-kambas is still with us. Sue Savage-Rumbaugh tells the following story about the arrival of a new animal, Pancho, supposedly a bonobo, at the Institute for Primate Studies in Oklahoma:[5]

> Therefore, even though Pancho was an adult male whom no one at the Institute knew, I felt confident in taking Pancho for walks around the Institute's farm, and I did most of my observations of him outdoors. We even took rides around Norman, as I did with Lucy, and we stopped for root beer, hamburgers, and fries at an A & W. Unlike being with Lucy, I never worried about Pancho grabbing anyone who approached the car, nor about him taking the steering wheel. Often I even took my young son along; Pancho was such a gentleman. Looking back, it was a crazy thing to do—and probably illegal—but I developed a very strong sensitivity to Pancho's skills and mine in our social and communicative interaction.
>
> After six to eight months we discovered that Pancho was not a bonobo after all, but a Koola-kamba, which some have suggested may be a naturally occurring chimpanzee–gorilla hybrid, or a chimpanzee–bonobo hybrid ... The fact that Pancho had been misidentified for so long is indicative of how few primatologists are familiar with bonobos. Had I realized Pancho's true identity, I'm sure I would have been less willing to be so free with him, thinking he might be aggressive (pp. 44–45).

Herr von Osten and Clever Hans

Only four years after Garner published his book, Herr von Osten gave the first public display of his famous horse, Hans. Hans tapped out his answers with his right hoof, and indicated readiness to answer a question by nodding his head affirmatively. He indicated that he was finished by tapping once with his left hoof. Hans II tapped out answers to mathematical questions, providing correct sums, differences, products, quotients, and even squares and square roots. Hans was so clever that that adjective was added to his name, whereupon he became "Clever Hans." He was clever enough to rather befuddle a scientific commission appointed to examine his abilities. The September Commission concluded that Hans answered questions without trickery, but did not think it possible at that time to reach any conclusion about how clever Hans was.

Hans was not clever enough to fool Oskar Pfungst, a psychology student assigned by Professor Carl Stumpf, a member of the commission, to investigate Hans' abilities more thoroughly. Pfungst discovered that Hans was not clever at all if no human who knew the answer was present where Hans could see him.

Pfungst believed that Hans used cues from observers who knew the answer. Pfungst's extensive and careful testing confirmed that Hans's success in answering questions depended on his ability to take advantage of changes in the posture of a human observer who knew when Hans had tapped the correct number of times. It was natural for observers to look down at Hans' hoof until the correct number had been reached, and then to look up. Another cue may have been postural relaxation when Hans had tapped out a correct number. Thus, it appeared that Hans was indeed very clever, but his cleverness lay in his use of subtle, unintentional cues rather than in understanding questions and knowing the answers to them. Since Pfungst's time, when situational cues rather than legitimate cognitive ability are used to answer questions, we call it the "Clever Hans effect." All students of animal language ability must be careful to eliminate any cues that might produce this effect.

The recent controversy about the use of "facilitated communication" by autistic children is a fascinating story related to this early era of animal language study. Claims were made that autistic children could type out messages (although they could not otherwise write or speak) if a facilitator held their hands while they pecked out their messages. In several cases, these facilitated messages accused a parent of sexual abuse. Two findings from the study of animal language cast great suspicion on the claims of those who believe in the genuineness of facilitated communication. First, the Clever Hans effect is not controlled for. It is therefore likely that the message comes from the facilitator, just as it came from those who were observing Hans. Second, in no case has an animal been more successful in spelling out messages than in writing (on a computer keyboard or with symbols) with words as units. Both these caveats are in addition to specific observations of facilitated communication in action, where it has been noted in many cases that the autistic child is not even looking at the keyboard while communicating.

Jacobson, Mulick, and Schwartz[7] reviewed the research on facilitated communication (FC) and concluded, in part, that there had been "a near universal failure of otherwise competent FC users to perform under controlled observation...." An awareness of animal language research, particularly of the Clever Hans phenomenon, might have protected people from a too-easy acceptance of the reality of facilitated communication with autistic children who cannot otherwise either read or write.[8]

However, those who claim that a Clever Hans effect accounts for the results in animal language experiments are responsible for demonstrating that they are correct, a point made forcefully by Premack and Ristau and Robbins. The latter authors had this to say about the problems associated with interpreting ape language experiments:[9]

> ... the researchers in the ape language projects are subject to expectancy effects, much like those reported from a long series of experiments by Rosenthal and other social scientists .... Such expectancy effects, state Umiker-Sebeok and Sebeok, will lead the researchers to observe and/or record ape behaviors inaccurately, to overinterpret ape performances, or unintentionally to modify the ape's behavior in the direction of the results desired (Umiker-Sebeok and Sebeok, 1980).
>
> To be sure, there are observational and recording inaccuracies, for instance, as indicated by Lyn Miles' measurement of intermethod agreement. Inaccuracies also often result from omissions of signs. Overinterpretation does occur, as we have discussed.
>
> For the most part, however, Sebeok asserts rather than proves effects due to Clever Hans errors. Inadvertent cuing through nonverbal signals can reasonably explain how a clever horse can know when to inhibit tapping his hoof, but contributes little towards understanding how an ape knows which particular finger and hand configuration to use in making a sign. More recent experimental work by Premack and Woodruff and by Savage-Rumbaugh and Rumbaugh and their associates have utilized, though to varying degrees, much stricter "blind" experimental designs in which the experimenter is out of the room during testing. In some experimental paradigms some nonverbal cuing still remains possible. We also wonder, though, if the chimpanzees are clever enough to discern such subtle cues, might they also not be capable of doing the experiment? All in all the effects of Clever Hans errors are powerful and the impetus to improving experimental design is a laudable one. It seems, though, that attributing the results of the ape language research to Clever Hans errors is unwarranted (pp. 238–239).

The analysis by Ristau and Robbins thus emphasizes the need for critics of animal language research to demonstrate that the results of experiments can be attributed to Clever Hans effects. However, equally, investigators of animal language must prove that Clever Hans effects are *not* responsible for the results. In short, the responsibility for proof is symmetrical with respect to the Clever Hans effect. Without proof in either direction, the cause for the results must be regarded as indeterminate. Because scientists are skeptical, a claim that an animal understands human language cannot be accepted unless other possible explanations of a performance have been eliminated. However, critics' explanations of animal ability by recourse to the Clever Hans effect when they have no direct evidence that nonlinguistic cues are responsible for the observed performance are as unscientific as an unproved claim that animals understand language.

William Furness and his Orangutans

One of the earliest attempts to teach great apes to speak was reported by William Henry Furness 3rd. Furness, after getting his A.B. at Harvard like his father and grandfather before him, got his M.D. degree from the University of Pennsylvania Medical School. Furness became a world traveler and something of an anthropologist, as well as a physician and surgeon. It was probably because of his travels to Borneo (he wrote a book on the headhunters there) that he became interested in orangutans. In 1909 Furness acquired both an orangutan (in South Borneo) and two chimpanzees, the latter from a trader in Liverpool. In 1911 Furness acquired another orangutan; he did not state its source.

In his report before the Philosophical Society, Furness did not describe the place or conditions under which he kept his animals. He does say that[10] "Frequently for weeks at a time I have spent as much as six hours a day in their company, but this is not one hundredth part enough" (p. 282). Modern investigators typically have someone with their ape subjects as much as 16 hours per day, and are probably much more consistent in their attention than Furness was able to be. In the context of discussing the excellence of his students' memories, Furness says[10] "After an absence of six months I have found that my apes have forgotten nothing that I have taught them, although during my absence their course of instruction ceased entirely and they refused to do for others what I had taught them" (p. 286). One rarely finds modern language researchers, with their more organized programs, allowing their participants to languish for such long periods of time, at least until some phase of their research has been completed. However, because great apes require such a great commitment of effort and resources, it is not unusual for a line of research to be dropped altogether. For example, when the Gardners lost their funding, they had to stop working directly with their animals and concentrate on the analysis of data that they had already collected; similarly, Herbert Terrace wearied of applying for financial support and sent his chimpanzee, Nim, back to Oklahoma.

Furness's most apt orangutan pupil learned to say "papa" after 6 months of training. The training included holding her lips together and releasing them while he repeated the sound, sometimes while in front of a mirror so that the orangutan could notice the similarity in movement between his lips and hers. One day, outside of lesson time, she said papa quite distinctly, and repeated it on command. She apparently connected the word with Dr. Furness, for she would, upon being asked, "Where's papa?" point to him or pat him on the shoulder. In another charming incident, when Furness started to carry her into the water, she became alarmed and said "Papa! Papa! Papa," while kissing him. After that he had not the heart to carry her farther into the pool. This incident may indicate that the orangutan, like the chimpanzee, has a tendency to vocalize primarily or only in emotion-producing situations.

Furness then taught the orangutan to say "Cup." That also took many trials, assisted by a spatula used to push her tongue back, and by Furness's placing his finger over her nose after she had taken a deep breath, so that she would have to breathe out through her mouth. That caused her to make the sound *ka*, which, after modeling and additional training, became a very clear *cup*.

Furness then progressed to training his student to make the sound *th*, preparatory to teaching her to say words like the, this and that. Unfortunately, the orangutan died soon after. Furness tried for 5 years to teach his surviving chimpanzee (one of the two died) to say "cup," but he had no luck in doing so. His conclusion:[10] "On the whole I should say that the orang holds out more promise as a conversationalist than does the chimpanzee; it is more patient, less excitable, and seems to take instruction more kindly" (p. 285).

In addition to contributing one of the early systematic attempts to teach apes to speak, Furness concludes his article with a wonderful quotation from

Reverend Sydney Smith:

> There may, perhaps, be more of rashness and ill-fated security in my opinion, than of magnanimity or liberality; but I confess I feel myself so much at ease about the superiority of mankind—I have such a marked and decided contempt for the understanding of every baboon I have yet seen—I feel so sure that the blue ape without a tail will never rival us in poetry, painting, and music—that I see no reason whatever, why justice may not be done to the few fragments of soul and tatters of understanding which they may really possess. I have, perhaps, felt a little uneasy at Exeter 'Change from contrasting the monkeys with the 'prentice boys who are teasing them; but a few pages of Locke, or a few lines of Milton, have always restored me to tranquility (p. 290).

Ape language researchers would feel very uncomfortable with the anthropocentric view of Reverend Smith. They would, however, probably agree that we should feel secure enough to do nonhuman animals justice about "the few fragments of soul and tatters of understanding" that they have manifested. Our challenge is to patch those fragments and tatters into as good a cloth of language as we, and they, can manage.

HUMAN LANGUAGE IN HOME-REARED ANIMALS

Toto and Maria Hoyt

Because Furness did not say how he cared for his pupils, it is not clear that his adventures with apes belong to a discussion of home-reared animals. Furness never married, so it is doubtful that the animals became members of his household. There is, however, no doubt about where Mrs. Hoyt cared for her gorilla, Toto; much of her book[11] describes how Toto did, and did not, fit into her household.

Mrs. Hoyt did not report any systematic efforts to teach human language to Toto, and Toto communicated to her only through natural gestures. However, it is nearly inconceivable that anyone who home-rears a great ape would not try to teach it language. No doubt Mrs. Hoyt and everyone else in her household talked in the presence of Toto, and often to her. In such an environment, human infants learn to speak their native languages without any need for heroic training. There is an age beyond which such exposure is unlikely to succeed in instilling full language competence in human children, but it is only in extreme circumstances that human children lack sufficient exposure to learn to speak and comprehend language. If we have learned anything from home-rearing observations, it is that there is a great gap between human tendencies to learn language and the tendency of other great apes to learn language, particularly on the production side.

Because the Hoyts did not perform systematic assessments, it is impossible to evaluate the level of Toto's comprehension of human language. It is clear that she did not produce spoken language. Mrs. Hoyt thought that Toto comprehended a good deal:[11] "We found that if one of us seeing her called to Tomas, [when Toto was running away from everyone] the result was only that she changed her direction and found a new hiding place, for although she could not talk she could understand Spanish as well as any Spanish child of her age" (p. 136).

Nadesha Kohts and Joni

Nadesha Kohts, like Mrs. Hoyt, did not report (at least in her English summary) any special effort to teach language to her male chimpanzee, Joni. Joni lived with Mrs. Kohts from the ages of $1\frac{1}{2}$ to 4 years, during the years 1913–1916. Kohts was struck by Joni's failure to vocalize, or even attempt to imitate human vocalizations:[12]

> The infant chimpanzee constantly hears human vocalizations, responds correctly to spoken directions, uses his own natural sounds for expressing his emotions and acquires complex conditional reflexes fot (sic) the mimetic expression of his desires. But never once has there been traced any evidence to the effect that the chimpanzee would try to imitate the human voice or to master be it even the most elementary words by means of which he would be able to greatly facilitate intercourse with his master (p. 576).

Kohts reported that Joni produced several sounds, "Almost all the 25 sounds that we have on record and which Joni was capable of emitting when moved by various emotional stimuli find themsleves (sic) a definite counterpart in Roody's vocalizations" (p. 575). Roody was Kohts' son, with whose behavior during the years 1925–1929 she compared Joni's earlier behavior.

Although Kohts has little to say about the difficulties of raising an infant chimpanzee, it is clear that her experiences did not lead her to have the "romantic" attitude toward her "baby" that Mrs. Hoyt had for Toto. Speaking of chimpanzees, Kohts says:

> ... the experiments conducted by Professor Yerkes have plainly shown that no amount of training will ever teach the chimpanzee ... human speech... the chimpanzee is devoid of imitation insofar as human sounds are concerned and generally fails to extend or improve its imitatory behavior ... the chimpanzee does not improve its motor habits connected with the use of tools or household implements ... he does not indulge in creative constructional play ... the chimpanzee ... actually fails to possess any inherent tendency towards progressing.... All the more strong is the contrast with the human child, who boldly dares to overcome his mental and physical deficiencies... It may even be ... the natural weakness of the human body—notably the weakness of teeth and arms—which prompted primeval man to toil, take a tool in hand and become technician and inventor (pp. 576–577).

Thus we see Kohts' romantic views deployed in favor of humankind, rather than in favor of chimpanzees! Both because I do not read Russian and because many years have elapsed, I do not know how much Kohts' writing was affected by the political climate in the USSR immediately prior to 1935. It may be that her admiration for humans who toil and "take a tool in hand" may have been more a product of Communist ideology than of comparative observations of Joni and Roody.

Gua, Donald, and the Kelloggs

The Kelloggs raised a chimpanzee named Gua for 9 months with their child, Donald. Their explicit goal was comparing the development of the two within the same environment. Donald was somewhat older than Gua, 10 months to her $7\frac{1}{2}$ months, at the beginning of the period of observation. Gua came from the Anthropoid Experiment Station of Yale University at Orange Park, Florida, although she had been born in captivity at the Abreu Colony in Cuba on November 15, 1930. The experiment began on about June 26, 1931, and ended on March 28, 1932. At that time[13] "... the ape was returned by a gradual habituating process to the more restricted life of the Experiment Station" (p. 276).

In many respects Gua, like Joni before her, compared very well with the human child. She walked, climbed, and ate with utensils as well as Donald. She made more errors during toilet training, but she was younger than Donald. Her hearing and sight were acute; as stated above, she was incredibly strong for her size, much stronger and quicker than Donald. However, again like Joni, she made no effort to imitate human speech, nor do the Kelloggs report that she said even one word. Their attempts to teach Gua arbitrary names for a small set of objects met with no success. Donald, however, did no better at this task, which adults regard as quite easy. Gua also failed, despite several months of training, to say the word "papa." The training consisted in placing Gua in the experimenter's lap, face up, while the experimenter said "pa-pa" slowly and distinctly. Later the experimenters manipulated Gua's lips in time with the utterance. The only result was that Gua, near the end of training, did occasionally twitch her lips.

Gua's limited articulation coincides with observations of chimpanzees by many other observers. The Kelloggs describe chimpanzee shortcomings in vocalization that may relate to acquiring human language:[13]

> There was no attempt on Gua's part to use her lips, tongue, teeth, and mouth cavity in the production of new utterances; while in the case of the human subject a continuous vocalized play was apparent from the earliest months. It was as if the child, like other normal humans of similar age, was *practicing* the formation of new vowels and consonants. In this manner the earlier cooing, singing, or humming of the young baby was transformed into a continual "la-la-la," "ngah-ngah," "gee-gee," etc., which constituted his later babble. Although Gua in her turn could form several vowels and although she seemed to be able to manipulate her lips and tongue with perhaps greater facility than the boy, no additional sounds were ever observed beyond those she already possessed when we first made her acquaintance. There were no random noises to compare with the baby's prattle or to the apparently spontaneous chatter of many birds. On the whole, it may be said that she never vocalized without some definite provocation, that is, without a clearly discernible external stimulus or cause. And in most cases this stimulus was obviously of an emotional character (p. 281).

This tie between emotion and vocalization was the subject of later comment by the Gardners, and the Hayeses said that Viki did not make a single voluntary sound between the time she stopped her occasional babbles at the end of 4 months and the beginning of her speech training at 5 months. There is no clear reason that vocalization in the chimpanzee should be so tied to emotional stimulation, nor is it clear that the tie is unbreakable. Viki was able, after much training, to make sounds voluntarily, and the bonobo, Kanzi, seems to have acquired this ability spontaneously.

Despite her shortcomings in vocalization, Gua did apparently understand 58 spoken phrases, compared to Donald's 68. For the first 4 months of their comparison, Gua was ahead of Donald. After that, Donald surpassed Gua in both the comprehension and production of human language, and would have surpassed her far more as time passed. Because neither ape nor child was tested in a context-free situation, we cannot know how much their apparent comprehension depended upon language per se and how much depended upon context and mode of articulation. Neither can we know, as the Kelloggs point out, whether phrases were responded to as combinations of words or just as perceptual gestalts that could be distinguished. They made no claim that Gua had true language ability; she could have been responding to nonlinguistic cues, as well as to the commands as perceptual gestalts.

Gua used nine gestures to indicate various desires, such as the desire to eat, drink, or sleep. Because Gua made spontaneous use of gestures and communicated through action in many nonverbal ways, the Kelloggs raised a question that has now been settled pretty definitively by the Gardners and others:[14] "The spontaneous use of gesture movements by chimpanzees raises the question whether this ability to gesture can be developed into something more. Could an intelligent animal learn a series of regular or standardized signals— as a sort of semaphore system? ... (p. 67). This is the question which has recently been asked by the Gardners."

Viki in the Hayes Household

A charming little book[15] by Catherine Hayes details the first 3 years of Viki the chimpanzee's life in the Hayes household. Those years sound like a disaster for the Hayeses, but nice for Viki; she was incredibly energetic and destructive. Viki babbled occasionally for the first 4 months and then stopped. Her vocal training began at 5 months, one month after her babbling stopped. The Hayeses had to shape her lips artificially in order to get her to say "mama," and before that they had to teach her for 5 weeks to make *any* sound voluntarily. (It seems that everyone who has tried to teach vocal language to apes had recourse to special lip training!) The Hayeses tried to use the techniques of articulation training as developed with humans. Later testing showed that "mama" was voiced, *papa* and *cup* not, but all sounds were made on expiration. Viki's natural tendency was to try to make sounds on inspiration. At 3 years she could say three words—*mama, papa,* and *cup*—and seemed to use *cup* fairly reliably, with little indication that she knew what the other two words meant. She also used *aaah* as a request, especially for a cigarette, and a clicking sound for a car-ride request. Now, 50 years later, it is difficult to imagine giving an infant chimpanzee a cigarette, but standards and beliefs about animal (and human) treatment have changed greatly in the interim.

Viki's understanding was spotty and uncertain, and the Hayeses seem not to have tested her comprehension very systematically. They were too aware of the need for careful testing to make claims about her ability to understand English without contextual cues. *Mama* tended to be used whenever there was a problem and Viki wanted attention. Viki also used *cup* with some reliability, but the use of *papa* did not indicate good comprehension. Viki's limited comprehension was certainly not related to lack of exposure to the use of language; English was constantly spoken around her.

Viki was a talented imitator, and could do delayed as well as immediate imitation; she dabbed at spots on Cathy's dress the next morning, after seeing Cathy do it the night before in order to clean up a milk spill. She sanded the furniture for hours after seeing the Hayeses sand a stake outside, and manifested many other cases of imitative behavior. It appears that Viki had a greater tendency toward imitation than Kohts' chimpanzee Joni, perhaps because Viki was much younger than Joni when she was taken into a human household.

Given her propensity for imitating motor behaviors, Viki's failure to imitate vocalizations is especially striking. However, it is consistent with Kohts' report that her chimpanzee, Joni, showed no tendency to imitate human vocalizations. Chimpanzees' greater propensity for imitating non-vocal motor responses may be related to the fact that these generally silent creatures are so much more generally active in the nonvocal arena. Future researchers might be able to elicit vocal imitation by beginning very early to encourage vocalization, which could lead to later attempts to imitate human vocalizations. Dr. Sue Savage-Rumbaugh reports (personal communication) that Kanzi often tries to imitate human vocalizations, even though he did not receive such special training. He vocalizes an easily comprehensible version of "right now"; like a human child, he shows a fondness for echoing his caretakers' most frequent command!

Returning to Viki, one interesting episode concerned an "imaginary pulltoy" that Viki appeared to be pulling around the house. Viki seemed somewhat embarrassed when Cathy Hayes "entered her world" by pretending to see and manipulate the imaginary toy, and added sound effects. On the first day, Viki accepted Cathy's joining the imaginary game. However, on the second day Cathy added sound effects that apparently frightened Viki, and the imaginary pulltoy game ended, never to appear again.

In her book Cathy presents a picture of Viki apparently admiring herself in a mirror. Very little is made of this, and we cannot be certain that Viki recognized herself in the mirror. However, many years later, Gallup developed a research industry around mirror recognition as a test of self-awareness; this apparently (to humans) simple perceptual/cognitive task is rarely if ever passed by untrained animals, and is passed by animals in only a few species even after training. Chimpanzees have so far been best at self-recognition, and they are not always successful.

Lucy, the Temerlins, and Janis Carter

So far as we know, the latest extended attempt by professionals to rear a chimpanzee from near birth exclusively in a human home was undertaken by the Temerlins, Maurice and Jane. The Temerlins must have been fans of Charles Schulz, not only because they named their female chimpanzee Lucy, but also because they earlier had a male chimpanzee whom they named Charlie Brown. Charlie met a tragic fate much worse than having a football withdrawn at the last moment before kicking; he suffocated himself in a blanket that he had looped around the bars of the ceiling of his crib. Although his loss was a psychological disaster for the Temerlins, they adopted Lucy at the age of 2 days not long after Charlie Brown died, and reared her in their home until she reached adulthood. Maurice Temerlin's book *Lucy: Growing up Human*[16] details the joys and tribulations of Lucy and her foster parents for the first 10 years of their lives together.

Lucy's history was quite different from that of previous chimpanzees raised in human homes. For one thing, the Temerlins did not regard Lucy as a subject in a scientific experiment; they regarded her as a daughter rather than as an object to be observed. The people responsible for her scientific career, such as it was, were not her parents; they were Roger Fouts and his students, who started to teach Lucy American Sign Language (ASL) when she was 5 years old. Among the graduate students who worked with Lucy was Sue Savage (now Sue Savage-Rumbaugh).

Because Lucy acquired a vocabulary of 100 signs, we can reach one conclusion immediately: chimpanzees can learn some sign language even after they are 5 years old. It becomes a reasonable guess, in the absence of proof, that the sensitive period for learning language ends later for sign languages than for vocal languages. Possibly the success in teaching language to feral children, which characteristically was quite limited, would have been greater had sign language been used.

Roger Fouts trained Lucy in ASL beginning in 1969 or 1970, during the period immediately after his arrival in Oklahoma. A short time before that Fouts had received his degree from the University of Nevada at Reno under the tutelage of the Gardners. Lucy's use of sign language revealed several interesting phenomena. For example, Fouts reported a case of attempted lying, involving Lucy's defecation on the rug[16] (the translations of Lucy's signs are in caps):

Roger: What's that?
Lucy: LUCY NOT KNOW.
Roger: You do know. What's that?
Lucy: DIRTY, DIRTY.
Roger: Whose dirty, dirty?
Lucy: SUE'S.
Roger: It's not Sue's. Whose is it?
Lucy: ROGER'S!
Roger: No! It's not Roger's. Whose is it?
Lucy: LUCY DIRTY, DIRTY. SORRY LUCY.

Although the glosses for the signs are probably generous (could Lucy really sign "Sue's," or did she just sign "Sue?"), it appears that Lucy was trying to avoid responsibility. She was being deceptive, and the ability to deceive is regarded as an advanced cognitive skill. It involves the attempt to manipulate the beliefs of the listener, and thus, at least in a sense, requires a theory of mind on the part of the deceiver.

Temerlin[16] reported that Lucy also constructed unique combinations of signs when she first encountered radishes and watermelons, for which she had no sign. Upon biting the radish, she signed "Cry hurt fruit"; the watermelon she called "candy drink" or "drink fruit." This is reminiscent of Washoe's creative use of WATER BIRD upon seeing a duck.

Lucy's third remarkable performance was her invention of a sign for "leash," which she indicated by placing a crooked forefinger near her neck. Again, Washoe behaved similarly, inventing a sign for "bib." Washoe's sign involved moving the fingers of both hands along her chest in the approximate position of the bib when present. Both Lucy and Washoe therefore invented iconic signs; many signs probably had iconic origins, which then tended toward increasing simplicity and abstractness. This suggests that a technique for teaching the use of signs, lexigrams, or other symbolic representations might begin, where possible, with iconic representations, and move gradually toward the conventional representation.

Lucy, like nearly all chimpanzees, became more difficult to manage as she aged. Temerlin[16] describes one example involving Lucy's relationship to Sue Savage:

> Sue was one of the brightest of the people who studied Lucy and was generally recognized as the top graduate student in the psychology department. For several years she had a beautiful, loving relationship with Lucy. Lucy was so affectionate toward her so consistently I would not hesitate to call it love. Then, after Lucy started menstruating she became more irascible and Sue could not control her. They began to have power struggles when they went outside together and Sue decided to end the relationship for the time being. I thought it was a great tragedy, but I could understand Sue's action. She had once lost part of her finger to an adult chimpanzee in a colony and hesitated to risk additional injury (p. 166).

One might recall in this connection that eleven fingers and a thumb were sacrificed to chimpanzee study during the history of the Institute for Primate Studies at the University of Oklahoma, and the fact that Premack and his students had to work with their star chimpanzee, Sarah, through the bars of a cage because her trainers otherwise had to spend so much of their time in avoiding bites. It is certainly understandable that chimpanzees are seldom, if ever, kept past adulthood in human homes.

As Lucy became more and more difficult to manage, the Temerlins had to decide what to do with her. They loved Lucy, but wanted to regain some control over their own lives. The last part of Temerlin's book is devoted to consideration of alternative futures for Lucy outside of the Temerlin household; could she be placed in a chimpanzee colony, or in a research setting, or returned to a wilderness for which she had become completely unadapted?

Part of Temerlin's description of a typical day in Lucy's life makes it clear that it would not be easy for her to adapt to a more typical chimpanzee life style:[16]

> Jane came home about 5:00 and I saw my last client at 6:00. Then the three of us had a gin and tonic together sitting in the living room. Then Lucy invited Jane and me to chase her... Nanuq and Lucy played chase ...
>
> Lucy got hungry while Jane and I were having a second drink, went to the refrigerator and helped herself to a carton of raspberry yogurt, a few bites of left-over pot roast, a carrot, half a carton of partially-defrosted frozen strawberries, and took three or four bites out of a head of lettuce. Then she went back to the sofa, covered herself with her blanket, and fell asleep (p. 210).

Linden picks up the story from there:[17] "Lucy was 11 years old in 1975, and the Temerlin's house and furnishings offered about the same resistance to Lucy that a house of balsa wood might offer to an exuberant human adolescent. Moreover, Lucy had reached an age at which it was unlikely that any new person could establish dominance over her" (p. 77).

The Temerlins were, however, fortunate. A young woman named Janis Carter arrived in Oklahoma when the Temerlins were considering separation from Lucy, and Janis was able to develop a good relationship with Lucy, despite the fact that Lucy was so old as to make that unlikely. Janis was also quite idealistic, and had objected to the constraints on the chimpanzees at the Institute for Primate Studies. At about the same time, the Temerlins heard about the work of Stella Brewer, who was rehabilitating chimpanzees to the wild in The Gambia, a very small African country. Stella was unwilling to take on Lucy, whom she regarded as a bad risk because she was an adult and, for a chimpanzee, extremely spoiled.

Despite reservations on the part of everyone involved, it was agreed that the Temerlins and Janis would accompany Lucy to Brewer's compound in The Gambia, and Janis would stay for a few weeks to help Lucy adapt. And so they went, in September 1977.

Janis Carter's stay had stretched from a few weeks to 6 years and counting when Linden wrote his book. The obstacles to Lucy's rehabilitation turned out to be enormous; they included both Lucy's own resistance and a variety of "political" difficulties, ranging from Janis's relationship with the Brewers to relationships with officials in The Gambia. Lucy nearly died on several occasions; in 1983 Jim Mahoney of Laboratory for Experimental Medicine and Surgery (LEMSIP) visited her and believes that, had he not been there to treat her hookworm, she would have died. Later, she did die, probably at the hands of a poacher. Roger Fouts[18] summed up Lucy's life and death in two poignant sentences. "Humans raised Lucy, taught her language, sent her to Africa, and rehabilitated her. And, in the end, they killed her" (p. 331).

Ironically, we might say that Janis Carter turned out to be the one rehabilitated; she replaced the Brewers as director of the rehabilitation center, and remains in that position as of 2002.

The present book is not the place to present the details of this epic, for which the interested reader can consult Linden. However, two points of immediate interest for animal language research should be noted.

First, Lucy did not stop using sign language, although Janis Carter ceased either using it or responding to it because she believed that the use of sign language might be a barrier to Lucy's rehabilitation. Janis relaxed her proscription upon the occasion of Linden's visit, and asked Lucy "Who he?" At first Lucy signed "Roger," and then, repeatedly, "Lucy Janis," "Lucy Janis." The only names in Lucy's original vocabulary were Barbara, Lucy, Maury, Roger, Steve (the Temerlins' son), and Sue Savage, so Lucy could never have answered Janis's question correctly. "Roger" was arguably the best approximation, the only "non-family" male for whom she had a name.

WHAT WILL HAPPEN TO "USED" CHIMPANZEES?

Second, part of Lucy's story has been and will be repeated for every chimpanzee student of language; that is, there will come a time when the chimpanzees' caretaker will face a decision about the animal's future. Washoe and the other Gardner chimpanzees have, so far, been fortunate; in 1970 they left Reno and went to Oklahoma with the Foutses. Later, when Roger Fouts broke with William Lemmon, the Director of the Institute for Primate studies, the Foutses moved them out of Oklahoma and found them a home at Central Washington University. In 1993 they moved into a comfortable facility including a large outside enclosure where they could run and climb. But what will happen if the Foutses lose financial support or, worse, die in an accident? What would happen to the chimpanzees then?

That is a worrisome question, even considering that Koko, Washoe, and Kanzi are the best known of all signing animals. Less famous chimpanzees like Booee have fared less well; he was moved to LEMSIP, where he is supposed to earn his keep as a participant in medical experiments. But Booee did not do as badly as some; his care was supervised by Jim Mahoney, the empathetic veterinarian. The two had the following exchange when Jim visited Booee:

> **Booee** (as usual): BOOEE WANT TREAT.
> **Mahoney**: (fearing that he did not have enough for all the animals in the room) "I don't have any" (in sign and words, pointing to his pocket).
> **Booee**: BOOEE WANT TREAT.
> **Mahoney**: I don't have any.
> **Booee**: (twice in rapid succession) BOOEE SEE (pointing to eye) SWEET IN POCKET.

This exchange shows that chimpanzees can catch humans in lies, as well as vice versa, as when Fouts caught Lucy lying about her feces. It also indicates that Booee was probably doing rather well, having both human and chimpanzee interaction in a healthy and reasonably uncomplicated setting; it appears that he is doing better than Lucy did in the wilderness, and he, unlike Lucy, is occasionally able to use the sign language that he went to such pains to learn. He is probably better off than Ally, who has apparently lost his human contact, as well as his name. A recent article in the *Los Angeles Times* reports that Mahoney has arranged to move Booee to a kind of chimpanzee haven in California, so his good luck seems to be continuing.

So what do we do with our old, used chimpanzees when we are through with them? We cannot cast them aside. Every chimpanzee removed from the wild is owed a lifetime of care. Every chimpanzee anywhere who is taught sign language deserves to be allowed to use it for life; and we should not forget that chimpanzees may live to be 60 or even 70 years old, if they are cared for well. Roger Fouts[18] discusses at length the moral dilemmas connected with zoos, medical research, language research and the fates of both wild and captive chimpanzees (pp. 388 ff). There are no easy answers, but we are moving toward more humane answers—and "humane" should apply to the apes as it does to humans. The federal government is making progress toward providing retirement facilities for chimpanzees that have been retired from research, and we hope they will soon be ready to accommodate all of them.

REFERENCES

1. A. T. Cory, *The hieroglyphics of Horapollo Nilous* (William Pickering, London, 1840).
2. A. C. Fraser. *An essay concerning human understanding, by John Locke* Oxford: (Oxford University Press, 1894), Vol. 1.
3. R. L. Garner, *The speech of monkeys* (Heinemann, London, 1892).
4. D. K. Candland, *Feral children and clever animals* (Oxford University Press, NY, 1993).
5. E. S. Savage-Rumbaugh and R. Lewin, *Kanzi* (Wiley, NY, 1994).
6. R. L. Garner, *Gorillas and chimpanzees* (Osgood, McIlvane & Co, London, 1896).
7. J. W. Jacobson, J. A. Mulick, and A. A. Schwartz, A history of facilitated communication: Science, pseudoscience, and antiscience. *American Psychologist*, **50**(9), 750–765 (1995).
8. B. J. Gorman, Facilitated communication in America: Nine years and counting. *Skeptic*, **6**(3), 64–71 (1998).
9. C. A. Ristau and D. Robbins, Language in the great apes: A critical review, in: *Advances in the study of behavior*, edited by J. S. Rosenblatt, R. A. Hinde, C. Beer, and M.-C. Busnel (Academic Press, NY, 1982), pp. 141–255.
10. W. H. Furness, Observations on the mentality of chimpanzees and orangutans. *Proceedings of the American Philosophical Society*, **55**: 281–290 (1916).
11. A. Maria Hoyt, *Toto and I: A gorilla in the family* (Lippincott, Philadelphia, 1941).
12. N. Kohts, *Infant ape and human child*, 2 vols. Scientific Memoirs of the Museum Darwinianum, Moscow (1935).
13. W. N. Kellogg and L. A. Kellogg, *The ape and the child* (McGraw-Hill, NY, 1933).
14. Winthrop N. Kellogg, Communication and language in the home-raised chimpanzee, in: *Speaking of apes: A critical anthology of two-way communication with man,* edited by T. A. Sebeok and J. Umiker-Sebeok (Plenum, NY, 1980), pp. 61–70.
15. C. Hayes, *The ape in our house* (Harper, NY, 1951).
16. M. K. Temerlin, *Lucy: Growing up human* (Science and Behavior Books, Palo Alto, CA, 1975).
17. E. Linden, *Silent partners* (Times Books, NY, 1986).
18. R. Fouts with S. T. Mills, *Next of kin* (Morrow, NY, 1997).

CHAPTER FIVE

Washoe, the First Signing Chimpanzee

WASHOE COMES TO RENO

It was June 21, 1966. Allen and Beatrix ("Trixie") Gardner welcomed a chimpanzee named Kathy to Reno, Nevada. They renamed their wild-caught infant chimpanzee Washoe after the county in which Reno is located. In an ironic touch, the Gardners and Roger Fouts later discovered that Washoe, in the language of the Washoe Indians who were the county's original inhabitants, meant "people,"[1] in the sense of "the people," or the special people of the region. Washoe was, the Gardners estimated, about 10 months old, and they intended to teach her a simple version of American Sign Language (ASL).

The Gardners' cross-fostered Washoe in an environment as close as practicable to one that would be provided to a human child. Washoe had her own homey space in a used house trailer parked in the Gardners' backyard. It had the furniture, kitchen, and toilet facilities that the previous owners had left behind. That kind of arrangement fulfilled the need for a human environment and constant attention to Washoe while allowing both the infant and her foster parents an independent life. Because she was treated like a human child, Washoe behaved much like a human child. Washoe always had human companionship during her waking hours, and help was as close as the intercom during the time that she was supposed to be sleeping. Even when there was no reason to expect problems, someone came into the trailer each night to check on Washoe.

THE CHOICE OF SIGN LANGUAGE

The Gardners were familiar with attempts by Keith and Cathy Hayes, as well as by others, to teach or encourage apes, most often chimpanzees, to learn a vocal

Allen and Trixie Gardner share a meal and signs with the young Moja, one of the chimpanzees in the Gardner's second project, in the chimpanzees' comfortable kitchen. Photo courtesy of Allen Gardner.

language, with very little success. The Gardners decided to use the gestural language of the deaf in order to bypass the vocal apparatus. The Gardners thought that chimpanzee vocalizations were too closely tied to their emotional state to make vocal language a good bet. And sign language had another great advantage: deaf human children used sign language, so it would be possible to compare the acquisition of sign in chimpanzees with the acquisition of sign in human children.

This gives away the ending before the story has well begun, but the Gardners' decision to try ASL turned out to be dramatically productive. At the age of 51 months, Washoe used 132 or 133 signs. Later, Roger Fouts reported that when she was in his care her vocabulary grew to 170 signs. Compare that to Viki's misuse of four poorly pronounced words after 3 (ultimately 7) years of hard labor on the part of the Hayeses! Moja, a female chimpanzee who studied in the second phase of the Gardners' project, achieved a vocabulary of 168 signs when she was 80 months old. Washoe did not begin training until she was about a year old. Moja and Pili, another chimpanzee in the second phase, were in the presence of sign almost from birth, so they not only achieved larger vocabularies while under the Gardners' tutelage, but also signed much earlier, at the age of 3 months! In addition, Moja was combining signs when she was 6 months old!

Sign language has another great advantage over the alternative techniques that have been used to circumvent the vocal channel: animals can "speak" rapidly. The Gardners cite a comparison between the utterance rates of the bonobo Kanzi, who communicated via lexigrams, and of Dar and Tatu, who used signs, when each was about 4 years old. Kanzi averaged 10.2 lexigram utterances per hour. Videotape analysis showed that Tatu averaged 441 signs, and Dar averaged 479.[2] This comparison of rate of language production is a dramatic illustration of the relative convenience of signing. Another of its advantages is that it requires no equipment—no keyboards, lexigram boards, or plastic symbols, for example.

In evaluating these impressive figures, one should remember that the Gardners' pioneering effort was the *first* to demonstrate the production of any appreciable number of human words by animals. The Gardners also went well beyond earlier efforts in the rigor with which they demonstrated both the production and the comprehension of signs.

CHIMPANZEE CREATIVITY

Washoe started to combine signs very quickly after she had learned 8 or 10 individual signs.[2] The Gardners cite examples like YOU BLOW and YOU ME DRINK. That was exciting enough, but Washoe soon followed up with a demonstration that she could provide information that was not available from the context alone. She dropped a toy down a hole in her trailer wall while in the care of one of the assistants. When Allen Gardner returned she signed OPEN at the place in the wall where the toy had fallen. He understood her request, and fished out the toy.

The abilities to combine words into meaningful phrases and to convey new information are indeed impressive, but Washoe and the other chimpanzees surprised the Gardners and later the Foutses with their inventiveness. The Gardners knew of no sign for the bib that Washoe wore at mealtime, and used the sign NAPKIN. Washoe, however, invented her own sign, consisting of drawing the fingers of both hands down her chest where the edges of the bib were located. It turned out that Washoe's invented sign was almost exactly the sign used in ASL!

Washoe later created meaningful and novel combinations of signs. She called her potty chair a DIRTY GOOD and a swan WATER BIRD. Moja replicated

Washoe's creativity by calling a cigarette lighter a METAL HOT and a thermos jug a METAL CUP DRINK.

Thus the Gardners' chimpanzees were not limited to what they were taught explicitly, and used language in their own unique ways. Perhaps the best summary of these accomplishments is the seven-page table[2] (pp. 304–310), listing the signs used by Washoe, Moja, Tatu, and Dar at the age of 5 years. The list, which is not exhaustive, includes 196 signs and the contexts in which they were used. Such a table is, however, at best a black and white representation of what is really a vividly colored and rich use of language by chimpanzees, the truly amazing outcome of the first attempt to teach apes a sign language.

THE GARDNERS' TRAINING METHODS

The Gardners entirely avoided the use of vocal English in the presence of their chimpanzees. Their rationale was that:[3]

"Attempting to speak good English while simultaneously signing good ASL is about as difficult as attempting to speak good English while simultaneously writing good Russian."

Often, teachers and other helping professionals who only learn to sign in order to communicate with deaf clients attempt to speak and sign simultaneously. Those who have only recently learned to sign soon find that they are speaking English sentences while adding the signs for a few of the key words in each sentence (Bonvillian, Nelson, & Charrow, 1976).

When a native speaker of English practices ASL in this way, the effect is roughly the same as practising Russian by speaking English sentences and saying some of the key words both in English and in Russian. It is obviously not a good way to master a foreign language.

It was clear from the start of Project Washoe that the human foster family would provide a poor model of sign language if they spoke and signed at the same time. Signing to the infant chimpanzee and speaking English among themselves would also have been inappropriate. That would have lowered the status of signs to something suitable only for nursery talk. In addition, Washoe would have lost the opportunity to observe adult models of conversation, and the human newcomers to sign language would have lost significant opportunities to practice and to learn from each other. Until 1986, the Reno laboratory was the only laboratory that maintained the rule of sign language only (pp. 6–7).

The Gardners' dedication to protecting Washoe from speech sometimes took amusing turns. They say[2] "We were always avoiding people who might speak to Washoe. On outings in the woods, we were as stealthy and cautious as Indian scouts. On drives in town, we wove through traffic like undercover agents" (p. 295).

Because there is little basis for comparison, we cannot say whether forbidding the use of vocal language was a good idea or not. However, it is clear that Project Washoe was a great success, and we have no wish to argue with success. Penny Patterson's procedures with her gorillas Koko and Michael included the use of both sign and vocal English; Roger and Deborah Fouts also sometimes combined English and sign. These projects were also very successful, but the available data are insufficient to decide the relative degrees of success. Because the projects vary in multiple ways and involve multiple species, it is impossible to ascertain how success was related to any single characteristic, with the use of vocal language as only one example.

The Gardners borrowed some ideas from the literature of operant conditioning, but soon returned them little used. Washoe's first signs were taught through gradual shaping and immediate reward, typical operant procedures. When Washoe approximated the sign for "more" (both hands placed together in front of the body or, in the infant version, overhead), she was rewarded with more tickling or more rides across the floor in a laundry basket. Analogous procedures were used for the sign "open" (holding the palms down and parallel and then lifting them apart). However, these operant procedures were too tedious, time-consuming, and ineffective for use with later signs. In discovering this, the Gardners reported that they rediscovered a lesson learned by Keith Hayes and Cathy Nissen[2] (Cathy Hayes reassumed her maiden name after a separation from Keith).

The lesson was that immediate reinforcement was often too distracting for practical application to the learning of signs. The Hayeses had discovered what they thought would be a very effective way to get Viki to communicate. They gave her a set of pictures, and Viki could let her needs be known by presenting the Hayeses with the appropriate picture; if she handed over a picture of an apple, she was rewarded with a piece of apple. However, Viki had become extremely fond of car rides, and she demanded immediate reinforcement for presenting a picture of a car. If she did not get it, she refused to attend to further learning tasks. The Hayeses removed the car picture from the set, but it was too late. Viki would not use the pictures, and she looked for pictures of cars in magazines, which she duly tore out and presented, becoming quite frustrated when she was no longer rewarded. This behavior is not unusual; animals rewarded for several kinds of activities have concentrated on the reward rather than on the activity.

The Gardners discovered early in the Washoe project that drill was counterproductive; their procedures thus contrast markedly with the later procedures of Herbert Terrace, who used systematic operant procedures with his chimpanzee Nim in an environment restricted in order to eliminate extraneous, distracting stimuli. These differences in procedure may have produced different results and different beliefs about the capacities of chimpanzees. I will discuss divergent opinions about chimpanzees' uses of signs after I have described Terrace's methods, observations, and conclusions.

The Gardners relied more on the techniques used by deaf parents in human families after they found that operant conditioning procedures were ineffective. They treated the chimpanzees like children, responding to their requests, answering their questions and drawing their attention to signs by signing on their bodies.

Two specific non-operant techniques are sometimes helpful in teaching sign language: modeling and molding. Modeling provides an opportunity for observational learning, which is expressed in immediate or delayed imitation. In order to maximize the learner's opportunity to learn from observation, parents, teachers and foster parents all model simplified language (called "motherese" in the case of vocal language) and repeat the infant's signs, often with amplification. Molding involves putting the learner's hands into the proper shape, which provides additional opportunity to observe, encourages imitation and provides proprioceptive and visual feedback related to the correct sign.

Roger Fouts and the other Gardner assistants played a crucial role in the move toward molding; molding is also called guidance. Although the Gardners initially discouraged its use, Roger and the others persisted because it seemed to work. The Gardners wisely encouraged innovations among their students, and they became convinced that guidance was useful, and even allowed Roger to do a dissertation comparing the effectiveness of guidance and other techniques in Washoe's acquisition of her signs. Amazingly, years later Washoe sometimes used molding to teach signs to her adopted son, Loulis!

Although observational learning and molding are often the techniques of choice, they do not always work. The Premacks unsuccessfully tried observational learning to teach their chimpanzees the artificial language of plastic symbols, but they had to revert to more systematic, operant-based techniques that required the animals to choose correct symbols and sequences of symbols in order to obtain rewards. Duane Rumbaugh used similar reward procedures

to teach his first chimpanzee student, Lana, to press correct sequences of keys. However, reward alone was insufficient; Lana failed to do anything without guidance from a human companion. It appears that different types of language tasks require different training methods; there is no panacea, no magic bullet for teaching either man or beast.

Shaping through the use of reward can succeed when the available responses or modes of responding are limited, as they are when only a few plastic symbols or keys to press are available. It is less successful when, as in the case of a free-moving chimpanzee, almost anything may happen! The experimenter in the latter case would have to wait a very long time before anything approximating the desired response emerged to be reinforced. The freedom from any apparatus besides the hands is a great advantage, and leads to much more spontaneous responding, but that same freedom makes it nearly impossible to teach signs through standard operant techniques.

It is important with human children, and perhaps even more important with young chimpanzees, to control motivation if they are to learn. Too little motivation and nothing is done. Too much motivation and attention is misdirected. The Gardners found it useful to assume, after their flirtation with operant techniques, that chimpanzees have an inherent motivation to communicate. Only occasional, moderate external rewards were needed to encourage their chimpanzees to use signs, and most interchanges were initiated spontaneously by the chimpanzee pupils.

The Gardners tested their beliefs about the obligatory nature of chimpanzee vocalizations by recording the responses of two of their pupils, Tatu and Dar, to events and to signed announcements presaging the events.[4] The chimpanzees were more likely to use a sign in anticipation of an event (the arrival of a person they liked, for example) than to use the sign when the event itself occurred. They were more likely to vocalize to the event itself than to the announcement, a finding consistent with the Gardners' belief that chimpanzee vocalizations are tied to the arousal of emotion. The vocalizations were similar to or identical with the vocalizations of wild chimpanzees, which is also consistent with the belief that in chimpanzees, vocalizations are obligatory, and probably have a dominant instinctive component. Finally, signing was, overall, far more frequent than vocalizations, even to the event itself. That is consistent with the tendency of chimpanzees to be relatively silent animals. Signing was less dependent than vocalizations on stimuli associated with emotional arousal. These observations vindicated the belief that the use of a gestural rather than a vocal language was consistent with chimpanzees' proclivities.

PROPERTIES OF CHIMPANZEE SIGNING

Like human children, Tatu and Dar often used reiteration and incorporation of added signs in response to positive events. It appears that these devices indicate agreement, assent or excitement with the message received. For example, Tatu responded to TIME ICECREAM NOW, by reiterating part of the message: ICECREAM ICECREAM ICECREAM seven times![4] Although Terrace, Petitto, Sanders, and Bever[5] stated that children rarely repeat words or phrases, the Gardners believe that this view is a mistake based on the elimination of such repetitions from some written reports, and that chimpanzees are similar to young children in this respect, as in other aspects of early language learning. A report by Nelson is cited[4] to support the Gardner position:

> It must be borne in mind that when researchers have analyzed children's utterances, as when in the past they have analyzed chimpanzees' utterances, repetitions do not count. This is true both for words and for whole sentences. Thus many of Nim's productions as reported [here] would not be recorded as 4, 5, or 16 word utterances but as 2 or 3 word utterances with repetitions. That repetitions don't count doesn't mean that they don't exist. My transcripts are full of false starts, repeated words, repeated phrases ... Anyone who claims that children don't repeat has led a sheltered library existence. (pp. 50–51)

The Gardners might also have cited the classic guide to style in the English language to argue that even human adults use repetition. In describing his teacher, Professor Strunk, White writes:[6]

> Will Strunk ... uttered every sentence three times. When he delivered his oration on brevity to the class, he leaned forward over his desk, grasped his coat lapels in his hands, and, in a husky, conspiratorial voice, said, "Rule Seventeen. Omit needless words! Omit needless words! Omit needless words!" (p. xiii)

We should not make too much of the parallel between Professor Strunk and Tatu, but Strunk's passion for eliminating words was expressed in exactly the same way as Tatu's passion for ICECREAM ICECREAM.

The Gardners are as careful about the descriptions of their signs as they are to note repetitions and incorporations. They provide a detailed accounting of how signs were recorded and tabulated for all of their chimpanzee participants.

The usual way to describe a sign is to specify the place (P) of the sign, the configuration (C) of the hand or hands, and the movement (M) involved in making the sign; the shorthand designation for this system is PCM. Nichols and the Gardners found it useful to supplement the standard descriptive system by adding the *orientation* of the hand and sometimes of the arm and wrist, the *contactor* (the part of the hand that contacted the place of the sign), and the *direction* of the movement (up, down, circular or toward or away from some landmark, like the lips). For example, Moja's 166th sign was glossed as the English *peach*. Its place of articulation was the cheek on the side opposite the signing hand, the starting configuration was an open or curved hand (whose orientation was with fingers pointing up), and whose movement was with fingertips rubbing down. This case is interesting because the place of articulation (the opposite cheek) differs from that for the standard sign (the cheek on the same side as the signing hand). The standard movement also involves closing the spread palm to a tapered hand while rubbing down.

There are many cases of variation from standard forms in the signs of chimpanzees, just as there are in very young human children during their acquisition of signs. Moja and Dar, both of whom moved the peach sign to the opposite cheek, may have found the changed sign easier to make because of their long arms. Chimpanzees, like young children, also often adapt signs by enlarging the scope of places where the signs occur; a popular example is the sign for "more," made by placing the fingertips together in front of the body by mature signers, but often made overhead by children and young chimpanzees.

Rimpau, Gardner, and Gardner[7] studied the variations in the forms of signs used by chimpanzees to find out whether chimpanzees, like humans, used such variations to modulate the meanings of the signs. An example of such inflections for the human case is repeating the "house" sign in order to convey the notion of "town." Another example is that the sign for "ask-question" means "I ask you" if it is directed toward the observer, but "you ask me" if directed toward the signer.

The authors reported on an analysis of eight samples of videotaped sign interchanges between Dar and one of his human companions, Tony McCorkle. Their question was whether chimpanzees really inflect their signs to change their meanings, as human signers do, or whether any variations in signs are essentially random errors in the shape, place or movement of the sign.

First, the authors found that variations were rather common. Inter-observer agreement about the identity of the sign and where it occurred was high, with independent observers agreeing on the appropriate English word (gloss) for the sign made by the chimpanzee about 80 to 85 percent of the time. Interestingly, agreement about the gloss for the human companions was no better (about 75 percent). In both cases, however, the agreement was regarded as satisfactory. This is similar to levels of agreement (71 to 92 percent) in a study of human children.

The variations were not random. There was a systematic relationship between the grammatical category into which the sign fell and the probability that the sign would occur in a modified form. For example, signs that were verbs either always (like "groom") or sometimes (like "brush," which can also be a noun) frequently occurred in modified form. Verbs were modified nearly two-thirds (62.9 percent) of the time! The most common type of modification was that the sign was moved to a non-standard location. A typical example would be that the sign for tickle was moved to a place on the signer's body (apparently the place where he or she would like the tickling to occur).

Another example of non-standard locations was on the observer's (Tony's) body. Dar placed signs on Tony's body more often when it was apparent that he wanted Tony to engage in some activity, or wanted to be sure that Tony was paying attention to his needs, specifically for his share of chewing gum, in which case he encouraged Tony to pay attention by placing the sign for gum on Tony's lips or cheek.

Human adults also sometimes move signs to non-standard places; an example taken[7] from Fischer and Gough (1978) is that "If one describes a mosquito bite, one shows where the victim has been bitten, and so on." Thus in varying the locations of signs, chimpanzees are manifesting yet another behavior shared with humans.

CHIMPANZEE SIGN VOCABULARIES

Dar, Moja, Tatu, or Washoe used a total of 430 signs;[8] one table describes the contexts of use and glosses for each sign and lists its users. The signs are grouped into broad classes. Proper names included the names of the chimpanzees and their caretakers, for example, Dar, Moja, Tatu, Washoe, the Gardners, Arlene K., etc. Generic names were boy, friend, and girl. The pronouns me, we, and you were included. Names for animate objects—baby, bird, and bug, for example—and inanimate objects, ranging from airplane to wristwatch, were commonplace. Many of the signs for inanimate objects stood for things that could be eaten. Apples, bananas, coffee, and ice cream are just a few of many instances. Some signs could function as either nouns or verbs, glossed as blow, brush, peek-a-boo, and vacuum, among others. Signs that functioned only as verbs were somewhat less in evidence, but bite, break, chase, laugh, and wrestle are examples of this group. Up and down, in and out, represent the locatives. Modifiers included colors, the materials of which objects were made, numbers, qualities other than colors and materials, and comparatives like same and different. Finally, the chimpanzees used several signs classified by the Gardners as markers and traits, including again, dirty, funny, help, sorry, and Washoe's first sign, more.

The particular forms of the signs are presented in another table of 81 pages. They were in many cases idiosyncratic, as is often the case with young human children. The individual ways in which the Gardners' chimpanzees formed the sign for flower are fairly representative of variations in the other signs. The standard form of the sign for flower is described as follows. The place of articulation is the nose. The configuration is "tapered hand, palm to signer." The movement is "fingertips contact one nostril after the other."

The Gardners' chimpanzees all approximated the sign for flower, but in no case did they duplicate it precisely. Washoe, Moja, Tatu, and Dar all placed the sign on the nose, as required. However, none of them used quite the correct configuration. Washoe's was described as "curved hand, palm to signer." Moja used an open hand with her palm toward her and fingers pointing up or to the side. Tatu used a pincer hand, and Dar sometimes used the same configuration, or sometimes the proper tapered hand, in either case with the palm sometimes down instead of toward himself.

The movement aspect of the sign also varied from individual to individual. Washoe sometimes contacted both nostrils as in the standard form, but not always. Moja behaved similarly. Tatu's index finger and thumb sometimes approached, sometimes touched, sometimes one nostril, and sometimes one after another. Dar usually repeated his contacts, and he sometimes made a sniffing sound along with the sign.

Thus the chimpanzees' signs tended to differ from the standard forms and from the signs of the other chimpanzees, but the signs were consistent enough within each individual and resembled the correct signs enough to be recognizable. Because chimpanzees are very active animals, it was a challenge to collect data on their signing and demonstrate that it was reliable. The Gardners and their students used videotaping and multiple observers to demonstrate reliability. It is obviously easier to check reliability and validity when a permanent record of behavior is available. Because the action can be slowed down or stopped, the place, configuration, and movement involved in a sign can be checked and rechecked. In addition, the fallibility of human memory for what happened is almost entirely overcome. Finally, it is easy to share videotapes with other researchers who may want to see the results for themselves. Studies of ape signing in the future will continue to rely heavily on videotapes for formal data analysis, but continuous recording is likely to remain impractical; thus unexpected events, like Washoe's request that Allen open the wall, will remain a source of new insights.

Despite all the caveats, the Gardner's success in increasing the size of the chimpanzee vocabulary by switching from vocal language to signs is stunning. Viki produced three or four vocalizations that resembled English, depending upon whether "up" is included. She also produced distinguishable non-English vocalizations to request cigarettes and car rides; probably only the latter two vocalizations were reliable enough to meet the criteria the Gardners established for accepting signs into a chimpanzee's vocabulary. Viki did not use "mama," "papa," "cup," and "up" reliably within the appropriate contexts. Within fewer than 5 years, fewer than the 7 devoted by the Hayeses to Viki, Washoe mastered 132 or 133 signs; that is slightly over a 28-fold increase. In addition, the Gardners produced evidence[9] that chimpanzees after 5 years of training were continuing to build vocabulary, and that their use of phrase types paralleled that of younger human children.

If a future researcher can give a chimpanzee five times Fouts's figure of 170 for Washoe, the chimpanzee will exceed the figure—800 words—usually cited as bestowing basic competence in the English language! And if the Gardners, the first generation of researchers trying the sign language approach, can achieve a 28-fold increase in vocabulary, who can be sure that a 5-fold advance is out of reach?

CHIMPANZEE UNDERSTANDING OF GRAMMATICAL CATEGORIES

One method of determining whether a human or an animal understands differing grammatical categories is to ask them so-called wh-questions

(questions beginning with who, what, when, why, or where). Questions asking how many? also belong in this category, despite not beginning with wh. The Gardners tested all of their chimpanzees at different ages with questions from the wh category, including "how many" questions, but excluding "when" and "why" questions. The latter are the last to develop in human children, and are probably too difficult for chimpanzees, at least those trained by methods devised so far.

To make quite a long story short, the chimpanzees manifested developmental trends similar to those of human children.[10] The most significant issue is whether the chimpanzees' responses to questions contained signs that matched the category required by the question. For example, a "who" question demands a proper name or a pronoun as the answer, not a location. A reply that would be both appropriate and correct if Tatu were queried with "What your name?" would be ME TATU. An answer that would be appropriate in that it came from the same grammatical category would be ME DAR. In evaluating chimpanzees' understanding of wh-questions and grammatical categories, appropriateness is more important than correctness. When Naomi Rhodes asked Tatu "Who me?" and Tatu answered MARTI, MARK, NAOMI, the first two answers, though wrong, were correct in the very important sense that they were proper names; Tatu might have forgotten the correct answer or, if we wanted to be generous, we might claim that she was joking. A tabulation of errors, like a tabulation of correct responses, shows that the errors were generally from the correct categories.[11] This is important because the errors, unlike the correct responses, could not have been learned by association, and thus had to be based on understanding of the category of the question.

The Gardner's chimpanzees' responses were therefore very significantly related to the category of the wh-question. However, the relationship was not perfect. For example, Moja at the age of 43 months responded with the appropriate category to questions about where some action was taking place in only 5 of 12 cases. When she was younger, at 32 months, she responded correctly to eight of eight "where" questions. By 74 months her performance on the "where" questions had recovered to 14 out of 19. In the "Who demonstrative" category, she responded to only four of eight questions at 32 months, but to 11 of 12 at 43 months. Such uneven development is not surprising, given the small number of observations possible at each age and the fluctuations in chimpanzee motivation.

Despite the unevenness of development and the imperfect performances of all the chimpanzees, it is perfectly clear that all of them had a tendency to pick their answers from the appropriate grammatical category. Van Cantfort, Gardner, and Gardner conclude[9] that "For all the subjects and all of the samples, the chi-square analyses were statistically significant" (p. 218). This is a strong, but not a perfect, result. A near-perfect result would have been if all the animals, *for all of the grammatical categories*, had responded at above-chance levels of correctness. A really perfect result would have occurred if every response for every animal were correct. However, that might be too much to expect for human adults, and certainly is too much for human children of the same ages as the chimpanzees.

A reasonable criterion for human children to demonstrate mastery of wh-questions is that they provide the appropriate category of answer about half the time. Washoe and the other chimpanzees met or exceeded this criterion for almost all the categories tested ("why and when" questions weren't asked) almost all the time. Washoe at the age of about 60 months furnished the appropriate category of answer to Who, What, Where, and Whose questions 84 percent of the time.

AN EVALUATION OF THE GARDNERS' ACCOMPLISHMENTS

We have already indicated that the Gardners' achievement was remarkable. It is worth reiterating that they were careful to set up clear criteria for accepting the signs used by Washoe and other chimpanzees as genuine. Only if a chimpanzee used a sign on 14 consecutive days was the sign recorded as mastered! When the Gardners began their work nobody would have predicted that critics would be so vocal and so eager to find flaws in their data and procedures, so the Gardners deserve special congratulations for following strict scientific procedures when they might have been content with a more naturalistic approach.

Even though the Gardners were unusually careful, critics have suggested that the Clever Hans effect could have accounted for the chimpanzee's responses. For the critics to be correct, the chimpanzees would have had to be more sophisticated than a whole array of experimenters who were doing blind testing—or the experimenters would have to be dishonest or deluded. Considering that the Gardners present evidence that the Clever Hans effect was excluded, and the critics have no evidence that it was not, the more reasonable course is to reject the critics' suggestion and proceed to a summary of their contributions.

A brief overview of these achievements includes the following. First, they taught their chimpanzees a striking number of signs. Second, they showed that the signs were modified in ways similar to the ways human children modify them, ways that relate to the locus of actions or the identity of actors or recipients of action. Third, they demonstrated that chimpanzees had a tendency to reply to questions with signs belonging to appropriate grammatical categories. Fourth, they proved that the chimpanzees could communicate information that was not known by the addressee, as in the defining incident with Washoe, who at the age of 27 months indicated that the wall should be opened so that she could retrieve a toy lost through a hole in the wall (such events became commonplace as Washoe grew older). Fifth, they found that operant techniques for teaching language were less appropriate in their setting than modified techniques (like modeling and molding) that were used to teach ASL to deaf children. This finding has affected all subsequent ape signing research. Sixth, they developed techniques for cross-fostering young chimpanzees that made their care pleasurable for human caretakers and enriching for the chimpanzees. Seventh, they developed techniques for gathering and evaluating data that were appropriate for signing apes or children, and thus made direct interspecies comparisons feasible. Eighth, they showed that the developmental course of chimpanzee signing paralleled in many ways the developmental course of the signing of deaf human children.

These individual accomplishments revivified the study of animal language. It had been in the doldrums for approximately 20 years following the Hayeses' work with Viki, which had met with such limited success. Penny Patterson started work with Koko as a direct result of hearing the Gardners describe their work. Roger Fouts studied with the Gardners at Reno, and would never have embarked on the study of animal language without their inspiration. Other researchers were influenced less directly; the Gardners provided Duane Rumbaugh with two kinds of impetus; first, they demonstrated that chimpanzees' linguistic capacity considerably exceeded what had been manifested by Viki; second, his interpretation of the difficulty of observing, recording, and evaluating chimpanzee signing persuaded him that a less labor-intensive technique should be tried, and inspired him to devise the first computerized system for teaching chimpanzees language. Finally, the Gardners influenced Sue Savage-Rumbaugh and Lyn Miles indirectly through Roger Fouts and Duane Rumbaugh. Thus the Gardners gave new life to the study of language in animals.

THE GARDNER TRADITION ENDURES

Although Beatrix Gardner died suddenly in 1995, the Gardners' work continues, often in cooperation with the Foutses. Two recent publications[13, 14] illustrate their continuing relationship. Typically, a student completes work on a master's degree at Central Washington University with the Foutses. Then the student transfers to the University of Nevada at Reno for doctoral work under the direction of Allen Gardner (Central Washington University does not offer a doctoral degree in psychology). In addition to his or her coursework at Reno, the student does research at the Fouts' facility, which in this circumstance becomes a kind of University of Nevada field station.

One of these cross-fostered doctoral students, Mary Lee Jensvold, studied conversations between chimpanzees and a human interlocutor.[13] She allowed the chimpanzees to start a conversation and responded with one of four kinds of probes: requests for more general information, on-topic questions, off-topic questions and negative statements. The chimpanzees' rejoinders were conversationally appropriate and similar to those of human children.

In a still more recent study[14] Mark Bodamer tested whether chimpanzees would respond appropriately to a human facing them or with his back turned while sitting at a desk. When his back was turned, the chimpanzees either left or generally made attention-getting sounds. When he was facing them, they signed immediately 98 percent of the time, and seldom made any sounds during the ensuing conversation. The chimpanzees' conversational styles were appropriate to those of the human interlocutor under experimental conditions, confirming a host of earlier observations. And the legacy thrives.

REFERENCES

1. R. Fouts with S. T. Mills, *Next of kin* (William Morrow & Co., NY, 1997).
2. R. A. Gardner and B. T. Gardner, *The structure of learning. From sign stimuli to sign language* (Erlbaum, Mahwah, NJ, 1998).
3. R. A. Gardner and B. T. Gardner, A cross-fostering laboratory, in: *Teaching sign language to chimpanzees*, edited by R. A. Gardner, B. T. Gardner, and T. E. Van Cantfort (State University of New York Press, Albany, NY, 1989), pp. 1–28.
4. R. A. Gardner, B. T. Gardner, and P. Drumm, Voiced and signed responses of cross-fostered chimpanzees, in: *Teaching sign language to chimpanzees*, edited by R. A. Gardner, B. T. Gardner, and T. E. Van Cantfort (State University of New York Press, Albany, NY, 1989), pp. 29–54.

5. H. S. Terrace, L. A. Petitto, R. J. Sanders, and T. G. Bever, Can an ape create a sentence? *Science*, **206**, 891–902 (1979).

6. W. Strunk and E. B. White, *The elements of style* (3rd ed.) (Macmillan, NY, 1979).

7. J. B. Rimpau, R. A. Gardner, and B. T. Gardner, Expression of person, place, and instrument in ASL utterances of children and chimpanzees, in: *Teaching sign language to chimpanzees*, edited by R. A. Gardner, B. T. Gardner, and T. E. Van Cantfort (State University of New York Press, Albany, NY, 1989), pp. 240–268.

8. R. A. Gardner, B. T. Gardner, and S. G. Nichols, The shapes and uses of signs in a cross-fostering laboratory, in: *Teaching sign language to chimpanzees*, edited by R. A. Gardner, B. T. Gardner, and T. E. Van Cantfort (State University of New York Press, Albany, NY, 1989), pp. 55–180.

9. B. T. Gardner and R. A. Gardner, Development of phrases in the early utterances of children and cross-fostered chimpanzees. *Human Evolution*, **13**, 161–188 (1998).

10. T. E. Van Cantfort, B. T. Gardner, and R. A. Gardner, Developmental trends in replies to Wh-questions by children and chimpanzees, in: *Teaching sign language to chimpanzees*, edited by R. A. Gardner, B. T. Gardner, and T. E. Van Cantfort (State University of New York Press, Albany, NY, 1989) pp. 198–239.

11. R. A. Gardner, T. E. Van Cantfort, and B. T. Gardner, Categorical replies to categorical questions. *American Journal of Psychology*, **105**, 27–57 (1992).

12. B. Gardner and R. Gardner, Two-way communication with an infant chimpanzee, in: *Behavior of nonhuman primates*, edited by A. Schrier and F. Stollnitz (Academic Press, NY, 1971), Vol. 4, pp. 117–184.

13. M. L. A. Jensvold and R. A. Gardner, Interactive use of sign language by cross-fostered chimpanzees (Pan troglodytes). *Journal of Comparative Psychology*, **114**(4), pp. 335–346 (2000).

14. M. D. Bodamar and R. A. Gardner, How cross-fostered chimpanzees (Pan troglodytes) initiate and maintain conversations. *Journal of Comparative Psychology*, **116**(1), pp. 12–26 (2002).

CHAPTER SIX

Signs in Oklahoma and Ellensburg

THE JOURNEY OF ROGER AND DEBORAH FOUTS

When Roger and Deborah Fouts were married in 1964 they could not have known about the journey they were about to undertake. They were students at Long Beach State, and both of them wanted to work with children. However, 2 years later Roger found himself being interviewed by Allen Gardner at the University of Nevada in Reno, as a potential research assistant to work with the young Washoe while Roger pursued his Ph.D. in psychology.[1] Roger was sure that the interview had been a disaster and that he had no chance to get the assistantship he had to have to survive—until Allen took him, as a kind of consolation prize, to see Washoe. She leapt into Roger's arms and gave him a giant hug, something that Roger never again saw her do with a stranger. Washoe had hired Roger! More than 35 years later, she still has not fired him. That central fact has directed the Foutses' lives.

Roger became a key player in the Gardner's work with Washoe in the fall of 1967. Until 1970, life for the Foutses was about standard for graduate students in psychology; that is, they were overwhelmed with responsibilities and stricken with poverty, while having their lives run by a domineering mentor, in this case Allen Gardner (not to overlook Washoe, who was equally domineering and far less inhibited in expressing her desires). Like many graduate students, Roger was making critical contributions to his mentor's task of teaching Washoe sign language.

In 1970 Washoe reached the trying age of 5 years, and the Gardners decided that it was time to move on to projects with other chimpanzees. Allen arranged for the Foutses to accompany Washoe to the Institute of Primate

87

Studies in Norman, Oklahoma, which was affiliated with the University of Oklahoma. They really had no choice, but anyway it appeared to be ideal:[1]

> The Institute was located in what sounded like an idyllic setting with lots of trees, a pond with three islands, plus housing for about twenty chimps and other primates... It sounded too good to be true. I would have a teaching job with a real salary. We could build a new home for our growing family, and Debbi could go to graduate school in Oklahoma when she was ready. (pp. 111–112)

Roger's statement that it sounded too good to be true turned out, as that statement often does, to be prophetic. The reason was the conflict between Roger and the director of the Institute, Dr. William Lemmon. Lemmon was domineering to both animals and people, using cattle prods on the former and other coercive techniques on the latter. Clearly, Lemmon polarized people; some held him in high regard, or he could not have achieved his position of prominence; he gave many people an opportunity to work with chimpanzees, for which they were grateful. Others who had to work under his direction had vehemently opposite opinions.

Fouts reports that Lemmon insisted that the chimpanzee cages had to be made of expanded metal, which has sharp edges around every opening, instead of the safe chain link that Roger had requested. As a result, Washoe's infant Sequoyah cut his toe on the metal; the cut became infected and contributed to Sequoyah's death a few days later.

Lemmon and Fouts responded very differently to misbehaving chimpanzees. Lemmon's "foster son" was named Pan. Pan once displayed at and spit on Lemmon. Lemmon then went to his house and got a pellet gun, with which he shot Pan until Pan submitted and threw himself on the floor.[1] At that point Lemmon made Pan spread-eagle himself against the cage wire while Lemmon dug out the pellets with a knife.

Fouts took a different approach when a bad-tempered chimpanzees, appropriately named Satan, threw feces at him as he passed by Satan's cage. Roger simply stood there while Satan pelted him with feces until it dribbled down his face and clothing. Then Satan "cleaned" him by repeatedly getting mouthfuls of water and spitting them on Roger. Roger still refused to flinch, and his strategy worked—Satan was nonplussed, gave a friendly hoot at Roger, and was friendly ever after! It is no wonder that people with such contrasting styles as Roger Fouts and William Lemmon could not form a lasting alliance.

As a final example, in 1978 Roger teamed up with the screenwriter Robert Towne to convince Warner Brothers Studio to build a chimpanzee island and birthing cage for Washoe in connection with filming *Greystoke: The Legend of Tarzan, Lord of the Apes*. As soon as Lemmon got wind of it, he insisted on taking over the whole Oklahoma part of the operation. Roger, excited that the chimpanzee might finally have a decent place to live, withdrew to the sidelines, but Lemmon offended Towne by whipping a horse while showing guests a potential location for the island, and Towne refused to work with him. The project was canceled.

Lemmon and the Institute were almost too much for Roger; the battles literally drove him to drink for a time, but he recovered and negotiated a move to a nearby facility that was not under Lemmon's control. It was not, however, a good place for chimpanzees, and Lemmon insisted that Ally, one of Roger's favorite chimpanzees, be returned to him. The Foutses decided that a move was necessary for the welfare of their chimpanzee family, which by early 1980 included Moja, whom the Gardners could no longer handle.

So Roger inquired at many places with no results, but he received a surprising offer from Central Washington University at Ellensburg, Washington. After much ambivalence, the Foutses decided to move there, and moved quietly away from Oklahoma in late August 1980. They and their growing chimpanzee family have been in Ellensburg ever since. At first their colony had to be housed on the third floor of the psychology building, hardly an ideal location for a chimpanzee family that by 1981 had five members.

And other aspects of life were not a bowl of cherries, even in Ellensburg. The University was not supplying enough money to support the chimpanzee family, which included Tatu and Dar, as well as Washoe, Loulis and Moja. The local community came to their rescue, donating money, time and leftover produce. Then in 1981, Warner Brothers took the *Greystoke* project out of mothballs and offered Roger $100,000 for consulting.[1] He joyously accepted, for the money was enough to pay his salary and support his chimpanzees for a year. Christopher Lambert, who played Tarzan in the movie, visited Ellensburg to learn about chimpanzees, particularly about Washoe, the model for his chimpanzee mother in the film. Roger trained all of the actors who were to play chimpanzees to walk like chimpanzees and to eliminate their stereotypes of chimpanzee behavior. Roger would not consult on a film that used live chimpanzees because of the abuses usually, or perhaps always, associated with the use of performing chimpanzees.[2,3] The film was a great success, but hard times were far from over for the Foutses.

Deborah and Roger Fouts communicate with Washoe, the grand old lady of chimpanzee sign language. They are in their new facility in Ellensburg, and Roger and Deborah are in the approved low position that chimpanzees prefer when interacting with humans. Photo courtesy of April Ottey.

However, one blessed event occurred on May 7, 1993, 13 years after they arrived in Ellensburg. After 15 years of planning and 10 of fund-raising, the Foutses' chimpanzee family moved to a grand new facility that allowed them to live indoors or out, at their choice. It appears that the hegira of the Foutses and their human and chimpanzee family has led to a place that is not too good to be true. It's just good, very good considering that the eastern slope of the Cascades is not exactly natural chimpanzee country.

A VOYAGE TO ELLENSBURG IN 2000

Going to Ellensburg does not involve a sea voyage, as the heading implies, but getting there from any major metropolitan area is not easy. From San Francisco, where I was when I made the trip, it requires a flight to Seattle, and a rental car for the last 120 miles east to Ellensburg. The flight from Atlanta, from whence Duane Rumbaugh came to join me, is even longer. The road from Seattle lies athwart the Cascade Mountains, and the scenery is as snowy-beautiful as you expect from mountain roads in February. Ellensburg itself is a big small town, with its period-piece old brick buildings that take you 100 years back to the early 20th century—really passe now that we survivors have squeaked across into the 21st century. So why would anyone go to all that trouble to get to Ellensburg?

The reason, the secret that you already know, is that the Foutses now have their five hairy animals living in a 2.3 million dollar, 10-year-old building on the north side of Central Washington University. These animals have names— Washoe, Tatu, Moja, Dar and Loulis—and they have names because the first four have been taught some American Sign Language by humans, and the fifth, Loulis, learned 55 signs from his adopted mother, Washoe and the three other signing chimpanzees. Sadly, Moja died on June 6, 2002, so now there are only the other four.

Roger and Deborah Fouts are the prime movers behind the facility, and thus the prime caretakers of the chimpanzees. Although the various subspecies of *Pan troglodytes* used to be called "common" chimpanzees, there's nothing common about these five, to whom thousands if not millions of hours of teaching and observation have been devoted. How many animals have provided the justification for spending that kind of time and that kind of money? Probably none; so they are really very uncommon chimpanzees. Washoe, Loulis, and Tatu are all *Pan troglodytes troglodytes*; Dar is a *Pan troglodytes verus*, and Moja had the fanciest species designation of all: *Pan troglodytes schweinfurthii*.

When Duane Rumbaugh, Ellen Levita, and I arrived at the facility promptly at 2:30 on the appointed afternoon, Roger Fouts greeted us. Roger still retains his farm-boy friendliness and openness—amazing characteristics when you hear that they have over 8,000 visitors a year, all of them, like us, no doubt falling over themselves in their eagerness to get a glimpse of these unique chimpanzees.

Upon recognizing our eagerness, Roger quickly ushered us into the observation room, where four students were observing through the layered inch-thick glass that separates humans and chimpanzees. The chimpanzees see people a lot more often than most people see chimpanzees, so they are not enthralled to see Duane or Ellen or me. They are, however, clearly pleased to see Roger, and Washoe is soon signing to Roger that she wants a shoe. He thinks she wants his shoe, but she indicates that she wants the shoe of one of the students. That of course she cannot have. So within a few minutes Washoe is working herself up to a display for the benefit of her new guests, and makes a mild rush at the window. Dar, who has been observing the whole scene, suddenly makes a dash and smash move against the window, striking a glancing but dramatic blow on the glass as he transits with extreme rapidity from our right to our left. That seems to satisfy the five hairy musketeers, and they exhibit no more displays while we are there. Soon Washoe, Moja and Dar are engaged in a three-way grooming session on a high shelf to our left, out of sight unless we approach the glass and look up. Duane says that in all his years of watching chimpanzees he has never seen more than three chimpanzees calmly grooming each other, and then it is usually a child with a mother who is engaged in grooming, rather than three adults together.

The central research questions being studied when we were there involved various aspects of conversational repair. (I described other dissertation projects in the previous chapter.) Repair becomes necessary when the listener fails to get the message, or at least fails to respond as the speaker requests. Deborah Fouts said that 26 research projects were underway. That is truly remarkable when you realize that no humans ever enter the chimpanzees' territory, so many types of experiments are precluded. Roger says that the animals are not "hairy test tubes," and they have not volunteered for any experimental manipulations. They certainly have not signed the consent forms required before humans can participate in an experiment. Thus the observers must stay passive and allow the chimpanzees to behave as they wish.

A second reason Roger cites is that the animals have to remember to be gentle when interacting with humans. He does not think they would attack humans, but they might get into a little family fight and injure humans who happened to be in the way. During the 20+ years that the Foutses have been in Ellensburg, no human has been injured by the chimpanzees. That is truly remarkable; chimpanzees are confirmed biters, and few researchers who have worked closely for a long time with chimpanzees have ten undamaged fingers.

I have asked every chimpanzee researcher I have ever met why chimpanzees bite. Nobody knows; possible reasons have been that they are defending their territory, that they are establishing dominance, that there is no one reason, or that they are jealous. Even Washoe, who is probably among the gentlest of chimpanzees, bit Karl Pribram's finger when he visited the Oklahoma facility. Pribram apparently contributed to the damage by pulling away and cutting his finger on the extruded metal cage that contributed to the death of Washoe's baby, Sequoyah. The fact that Washoe immediately signed "Sorry, sorry," may indicate that her bite, and perhaps some bites of other chimpanzees, was instinctive or unintentional.

CONTRIBUTIONS TO ANIMAL LANGUAGE RESEARCH BY THE FOUTSES

Roger Fouts as a graduate student helped to convince the Gardners that imitation and guidance, not shaping and reinforcement, were the key to training Washoe to use sign. Allen and Beatrix ("Trixie") Gardner originally intended to use primarily traditional operant techniques to condition Washoe to use sign. But Roger and the other assistants showed them that guidance, which the Gardners later named molding, procedures worked better. Four and a half decades of language research have vindicated this view.

The final proof is probably Loulis's acquisition of signs from the other chimpanzees, who have no conception of conditioning. Social interaction that provides the learner an opportunity to observe the production of and response to language, whether vocal or signed, appears to be critical. Humans used only seven signs in Loulis's presence: WHO, WHAT, WHERE, WHICH, WANT, SIGN, and NAME. Loulis learned 51 or 55 signs, depending on whether four homonyms were counted as one or two signs, from the other chimpanzees. None of the signs that Loulis used were among the seven used by humans in his presence. A few samples of his signs are ALAN, BIRD, BLANKET, BOOK, CHASE, COME/GIMME, DIRTY, DRINK, GUM, HUG, NUT, ROGER, TOOTHBRUSH, and WASHOE.[4] An amusing side effect of humans' refusal to use other signs when Loulis could see them was that, when humans later did begin to use other signs in his presence, he refused for 4 months to pay any attention to them.[1] Roger Fouts said that it was as if Loulis were saying, "That's my language, not yours."

The results of denying Loulis any human modeling in sign are significant, not only for the above reason, but also because they show the transmission of a chimpanzee signing culture. Loulis gives us a reason to believe that chimpanzees, once taught to use a simple sign language, will pass it on to succeeding generations that will continue to use it. That confirms the observation that wild chimpanzees differ in their cultures with respect to nut cracking, hunting, and other activities.

DEBORAH FOUTS DECISIVELY REFUTES
THE CLEVER HANS CLAIM

Deborah Fouts' thesis committee challenged her to prove that chimpanzee signing was not just another case of the Clever Hans effect. She responded by videotaping the interactions of their five signing chimpanzees when no humans were present. Under those conditions no Clever Hans effect could occur. The results were spectacular. The chimpanzees signed often to each other in the absence of human observers. Researchers who observed the videotapes agreed 90 percent of the time about the meaning of the chimpanzees' signs. Deborah's remote recording showed that Loulis used the following signs either in conversations with the other chimpanzees or in signing to himself: HURRY, BALL, ME MINE, HAT, GOOD, COME, MORE, DRINK, WANT, PEEKABOO, GUM, PERSON, GIMME, OUT, SHOE, FOOD, MASK, and TICKLE.[5]

Her procedure was so simple that it seems amazing that it had not been performed before; however, there are few if any other groups of signing chimpanzees anywhere in the world, and it is often the case that the most ingenious experiments look simple once they have been done! Deborah's observations should, but may not, be the final nail in the coffin of the claim that all chimpanzee use of sign language is a matter of the Clever Hans effect. It would be amusing to hear a critic claim that the chimpanzees were cuing *each other*!

None of this implies that chimpanzees do not use subtle cues derived from humans and other chimpanzees to direct their behavior. On the contrary, they are astute observers of nonverbal cues, and can give the impression of understanding language when they don't. Austin, one of Sue Savage-Rumbaugh's two most talented *Pan troglodytes*, was a master at seeming to understand English when he could see Sue's face. In a striking video, when Sue asked for the headphones that Austin was wearing and held out her hand, Austin immediately took them off and handed them to her. But when, just previously, he was questioned via those same headphones and could not see the speaker, his responses had been random.

Finally, the Foutses have provided for all people a model of true respect for chimpanzees. They no longer restrain their chimpanzees for experimental purposes, nor do they manipulate them in any way. Their research is a matter of observation of the natural behavior of chimpanzees. Of course their chimpanzees are no longer very natural, having spent thousands of hours in learning and using sign language. Nor is their habitat very natural, considering that they are necessarily restricted to a cage, large though their indoor and outdoor quarters are. Nevertheless, the Foutses are demonstrating how much can be learned through observations that are as naturalistic as they can be with captive chimpanzees. Deborah's videotaping study shows that such observations may even be more decisive than experiments, contrary to the traditional view!

IMPROVING THE LIVES OF CAPTIVE CHIMPANZEES

The Foutses have decided that advocating *for* chimpanzees is more important than experimenting *with* chimpanzees. Roger Fouts underwent what amounted to a religious conversion when he saw a videotape of and later visited the biomedical laboratory run by Sema, Inc., in Rockville, Maryland. The video, made by an organization called True Friends, portrayed living conditions for chimpanzees that were almost too hideous to be believable— solitary animals raised in tiny isolettes that they had no hope of leaving, even for a brief outing. The smaller isolettes were about 2 feet wide and $3\frac{1}{2}$ feet high. Some of the chimpanzees were infants, others adults; nearly all of their cages were so small that they could hardly stand or even turn around. Solitary confinement is hell for an animal as social as a chimpanzee, and most were infected with HIV or hepatitis.

Although Sema was clearly violating the federal government's own guidelines for the care of laboratory animals, there seemed to be almost no concern about that fact in the scientific community. Jane Goodall was the exception; she filed an affidavit complaining about conditions at Sema. The laboratory director chided her for going public without visiting the laboratory, and invited Roger and Jane to come for a visit. They went, and conditions were exactly as portrayed in the video. Juvenile chimpanzees were confined in some isolettes 26 inches wide by 31 inches deep by 40 inches high! When Jane was asked by a National Institute of Health official to write a letter confirming that the animals at Sema were well cared for and that the laboratory was up to the standards of the United States Department of Agriculture (USDA), she said "By no means will I write you any such letter."

And she did not. Instead, she published an expose of the Sema lab. Roger rededicated himself to improving the welfare of chimpanzees, at great personal cost because asking for humane treatment of laboratory animals made him *persona non grata* with all the organizations funding research. This is not the place to provide the details of the fight for animal rights, (see Fouts,[7] or Blum,[8] or Peterson and Goodall[9] for those details), but one pair of court rulings is too amazing to omit. Roger in 1991 joined in a suit against the United States Department of Agriculture, and in 1993 District Judge Charles Richey ruled in the plantiffs' favor and wrote "a stunning rebuke" to the government and the biomedical industry for their illegal treatment of animals. But the court of appeals overturned the ruling, not because the facts were wrong, but because humans had not suffered harm and therefore could not sue! Only the chimpanzees could sue, and of course they are, so far, not able to sue on their own behalf.

Roger's recent publications[10] reflect his dominant interest in the welfare of chimpanzees. Most appear in a book or journal on animal rights or in the *Friends of Washoe* publication, rather than in a traditional research journal. These publications have probably confirmed his opponents in USDA and the research establishment in their belief that the Foutses have become advocates rather than researchers, and should be opposed rather than supported.

All of the Foutses' efforts to obtain legally and morally reasonable treatment for animals meant that he and Deborah were on their own with respect to supporting their chimpanzee family. They could no longer expect funding from the federal agencies whose treatment of experimental animals they so rightly opposed. Their chimpanzees might literally have starved without the generosity of the people at Albertson's produce department, who allowed them to pick through their aging produce for chimpanzee food, and of many other people in Ellensburg who contributed food, money and help. One family allowed the Foutses and their helpers to pick 700 pounds of fruit!

Despite the grinding effort to keep the chimpanzees fed and housed, the Foutses continued to plan and work toward building a facility that would allow the chimpanzees to go outside and enjoy a more satisfying life. The Washington legislature finally agreed to pay 90 percent of the cost of a new $2.3 million facility, and the chimpanzees moved in on May 7, 1993. It was just in time for Moja and Tatu, who had developed rickets to a life-threatening degree because they had been almost completely deprived of sunlight for 11 years. The freedom to go outside almost miraculously restored their health within months. They stayed outside until their skins turned bright red. Although their diet had been fortified with vitamins, vitamin D is not converted into a form usable for bone growth without sunlight.

Unfortunately, the new facility did not mean the end of the Foutses' financial problems. According to the Foutses, Central Washington University had promised to provide $210,000 per year to fund the facility, but actually provided substantially less the first year and much less later. So the Foutses had to find new ways to fund themselves, and they did. They got support from their foundation, Friends of Washoe, and started a series of teaching events called Chimposia that allowed visitors to learn about chimpanzees by visiting the Foutses' Chimpanzee and Human Communication Institute, where they could hear lectures, watch videos and, if the chimpanzees were willing, see chimpanzees playing and interacting. The $10 visitors pay for tickets, and the donations from Friends of Washoe, continue to support the Institute. Their Foundation (gifts to it are tax deductible) is at the Chimpanzee & Human Communication Institute, Central Washington University, 400 East 8th Avenue, Ellensburg, WA 98926-7573. Their web site is at http://www.cwu.edu/~cwuchci/washoe_friends.html. The Foutses and their supporters deserve enthusiastic support, not only for contributions to animal language research, but also for courageous and unselfish devotion to the welfare of both wild and captive animals.

REFERENCES

1. R. Fouts, with S. Mills, *Next of kin* (William Morrow & Co., NY, 1997).
2. D. Blum, *The monkey wars* (Oxford University Press, 1994).
3. D. Peterson, and J. Goodall, *Visions of Caliban* (Houghton-Mifflin, Boston, 1993).
4. R. Fouts, Transmission of a human gestural language in a chimpanzee mother-infant relationship, in: *The ethological roots of culture*, edited by, R. A. Gardner, B. T. Gardner, B. Chiarelli, and F. X. Plooij (Kluwer Academic Publishers, Dordrecht, 1994), pp. 251–270.
5. D. Fouts, The use of remote video recordings to study the use of American Sign Language by chimpanzee when no humans are present, in: *The ethological roots of culture*, edited by, R. A. Gardner, B. T. Gardner, B. Chiarelli, and F. X. Plooij (Kluwer Academic Publishers, Dordrecht, 1994), pp. 271–284.
6. R. Fouts, (1998). On the psychological well-being of chimpanzees. *Journal of Applied Animal Welfare Science*, **1**(1), 65–73; R. Fouts (2001). The state of the planet of the apes. *Friends of Washoe*, **22**, 1, 3–5; R. Fouts (2002). Darwinian

reflections on our fellow apes. In B. Beck et al. (Eds), Great apes and humans: The ethics of coexistence. Washington and London: Smithsonian Institution Press (pp. 191–211).

7. R. Fouts with Mills (1997).
8. D. Blum (1994).
9. D. Peterson and J. Goodall (1993).
10. R. Fouts (1998, 2001, 2002).

CHAPTER SEVEN

Koko Fine Sign Gorilla

A VISIT TO KOKO, NDUME, PENNY, AND RON

On November 19, 2000, I visited Koko, her male consort Ndume, and her people, Drs. Francine (Penny) Patterson and Ronald (Ron) Cohn. The way to their place leads off highway 280 north of Palo Alto, California, through the small town of Woodside. From there the road winds tortuously up into the Santa Cruz Mountains, nearly, I judged, to the highest point that can be reached by road. As I climbed higher and higher I was thinking, "This may be an appropriate place for mountain gorillas, but Koko and her male consort Ndume, are lowland gorillas—and this certainly isn't lowland!"

Steve Wise, a lawyer who specializes in animal rights legislation and the author of the book *Rattling the Cage*,[1] was following me, and I was following Penny, who has been working with Koko for the past 30 years. Penny promised to drive slowly (she said with her precise grammar), and as we screeched around the corners I thanked God that she thought she was driving slowly. A few hundred feet after the road split off the main road to the left, she pulled off to the right into a short unmarked driveway interrupted by an iron gate with a padlock. It seemed a strange sort of place for what is probably the world's most famous nonhuman animal.

I parked just outside the gate and Steve parked just inside. We then took a path off through the woods to get to the office; Ndume's home was off to the left, and Penny said they did not want strangers to upset him. We learned later that Ndume had diarrhea and throws feces at visitors, which would have been another reason we could have accepted for giving him a wide berth. Because we heard no sounds from Ndume, we assume that he did not hear us go by.

The office is unpretentious, to say the least; it is a one-story remodeled poultry house. However, once inside, you get the feeling that you are in an Apple orchard; that is, there is a profusion of colorful iMacs, one for every one of the 8–10 desks in the office. Apple was a strong corporate supporter of the Gorilla Foundation until it fell upon evil times, and the Gorilla Foundation remains a loyal supporter of Apple. DeeAnn, a full-time employee (*really* full time; our visit was on a Sunday afternoon at about 2:00 p.m.) greeted Steve and me.

Steve, DeeAnn, and I were soon joined by Penny, Ron Cohn, Anthony (who was then about 8 years old) and the friendliest German shepherd I have ever seen. Penny left to settle Koko down (Penny had been gone for a week, and Koko gets a little upset in her absence). Steve and I exchanged information and anecdotes for a few minutes until she came back and said it was "time go see Koko." We were more than ready.

Koko's home is perhaps 75 feet from the office. The sliding glass door outside the sturdy wire mesh was open, and a narrow porch is just outside the doors. The porch is really more like a walkway, about $3\frac{1}{2}$ feet wide and running along beside the building for 8 or 10 feet. The day was clear and fairly sunny, but as we approached I could see nothing inside. Koko's coat is quite dark, and she was simply invisible until you got your eyes out of the sun and into the shade. The glare from the windows on the far side of the building made it almost impossible to take pictures of anything as dark as Koko, as I found out later when I tried to print mine.

As soon as my eyes were acclimated I saw Koko squatting (or sitting; I think for gorillas they are the same) just inside the open sliding glass door. Anthony, Steve, and I were eager to see Koko, so we crowded onto the porch. Penny pointed out that Koko liked people to be down at her level (so do bonobos and chimpanzees), so Steve sat on a short stool, and I sat down on a low window box that was sitting on the outside part of the porch. As I sat there, or squatted just outside the door, Koko's eyes and mine were on the same level. I got the impression that Koko was about my size, an impression that she corrected later.

At first I was afraid to touch Koko, and certainly was not about to put my fingers inside. I remembered too vividly the number of fingers lost at Lemmon's primate facility, Sue Savage-Rumbaugh's short finger and Karl Pribram being a finger short, all as a result of bites from our frenetic friends, the chimpanzees. Even Maria Hoyt's female gorilla Toto was a bad biter. But Koko immediately greeted me by pursing her lips, showing her tongue, and blowing through the wire mesh at me. I returned the greeting by blowing back—our lips were perhaps three inches apart—but then I thought, "What if

I give Koko a cold?" I turned to Ron Cohn, who was standing behind me, and he agreed that that would be a bad idea, so I stopped blowing. I learned later that blowing is Koko's standard greeting, and she expects you to blow back. Penny thinks it is Koko's way of finding out how you smell. I must have passed the odor test, for Koko seemed to like me. Koko's breath was warm and pure, a lot like the breath of a human child. I also touched her lips, which felt like black velvet.

Koko then put her fingers through the mesh, and I hesitantly rubbed her fingers and gave her a chance to, if she wished, mash my fingers against the mesh. She treated me as gently as if I were a baby. We repeatedly touched and rubbed fingers; her fingers have the feel of soft leather.

Throughout all of these interchanges, Penny was carrying on a sporadic conversation with Koko in sign and English. I do not know sign, so I cannot vouch for the accuracy of her translations, but everything that happened was exactly as I would have expected from Penny's translation. That is, if Penny reported that Koko wanted to PLAY CHASE with someone, Koko either chased down to the far end of her building just before or just after her chosen chase-mate did the same. When Penny said Koko wanted to play TUG, Koko placed the end of a long strip of blanket through the mesh, and Steve and I made futile and justifiably hesitant efforts to keep Koko from taking it away from us. Once I decided to really try to hold on and cheat by angling it around the wire mesh. Koko took off with her end, and when the slack was out of the cloth I might as well have been trying to hold back a tractor on the other end of the line. It whipped out of my hand instantly.

That awesome strength makes it all the more amazing that Koko is the gentlest being I can imagine. I have already said that she treated my fingers with great delicacy. Penny fixed Koko a warm snack of cut vegetables and fed them to her through the mesh. Koko took each small bit in her large but extremely dexterous fingers and ate with perfect patience and politeness. Penny said that Koko had signed SPICE. On one strip of vegetable Koko detected a tiny fleck of ginger, picked it off, and passed it back to Penny as proof that the food was spicy. I am not exaggerating when I say that her dexterity in doing that appeared to be beyond what I could have achieved.

In addition to tug and chase, Koko liked to play scare. The object of this game is for Koko to scare the hell out of you, which she does by racing down to the far end of the trailer, sometimes adorned with a rag over her head or shoulder, and then racing back and coming to a stop perilously close to the mesh in front of your face. Needless to say, this game brings about its desired effect very well, at least for people who are not yet onto Koko's incredible gentility. Your initial impression that Koko is about your own size disappears really quickly as she thunders along inside her rooms.

It appeared that Koko initiated and controlled almost every communicative interchange. She did not imitate Penny, and it was very clear that the interchanges were truly conversational, with conversational repair when Penny was not able to see at first what Koko was signing. On only one occasion did food enter the picture, just prior to Penny's heating Koko's vegetable snack. It may be that some apes' conversations are predominantly requests for food, but that is not true of Koko.

When we first arrived, Penny said that Koko wanted to know who we were. I had thought about that before we arrived, because I knew that signers often identify people by referring to some unique physical feature; a person with a long moustache might be signed by drawing the fingers of both hands across the upper lip, starting in the middle and ending where the moustache ends. So when Koko asked, I stuck out my upper teeth and showed the big gap I have between my middle front teeth. I thought I might become "gap tooth." But, according to Penny's explanation, Koko has a sign that is glossed FAKETOOTH that she uses for gold teeth, and she modulates it for an upper location in the mouth, so I became UPPER FAKETOOTH. I thought I had been favored by a novel combination of signs until I learned by reading Penny's newsletter, *Gorilla*, that both Fred Rogers of "Mr. Rogers' Neighborhood" and Penny's father had been honored by the same designation. I am still honored to be placed in that company, even though I have not earned it by having any gold teeth.

Steve used the same strategy as a result of Koko's asking, again according to Penny, about gold teeth. Steve has one, which he showed to Koko, and he became simple FAKETOOTH. After we were named, Koko used the names to order which playmate she wanted for chase, tug, and scare.

Ron and Penny patiently translated, explained, and just plain waited for over an hour and a half while Steve and I exulted in interacting with Koko. Finally we realized that we were imposing at an intolerable level, and started to say goodbye to Koko. However, Koko is fascinated with nipples and, through Penny, she asked to see ours. So first I unbuttoned my shirt and showed her mine; then Steve pulled up his sweater and showed her his. As he did so, he asked Penny why Koko liked nipples. I said, "Steve, why do you like nipples?"

He said, "You didn't ask me whether I liked nipples." I said, "I didn't have to." (I was on very safe ground here; Steve has one 8-year-old and twin 5-year-old children.) Penny asked Koko, in sign and English, why she liked nipples, and reported that Koko replied NIPPLES ARE NIPPLES. I thought that was exactly right.

We left soon after, and this time we walked within view of, but out of throwing range of, Ndume. Penny said we should do that because Ndume liked to see people leave. Ndume was high above us in the upper area of his cage, a magnificent silverback who appeared to be at the height of his powers, as well as at the height of his cage. At least when he does not have diarrhea. Diarrhea is not uncommon among male gorillas as a reaction to intruders, probably because their responsibility for protecting their families is stressful.

Steve and I then went to the offices of the Gorilla Foundation on Woodside Road in Redwood City, a few miles east of the language facility. There we were treated to a full explanation of the progress of the facility to be built on Maui, a much more lowland-gorilla-friendly location than the Santa Cruz Mountains. A 70-acre plot has been donated for a gorilla preserve, 12 acres are cleared for the first buildings, and some foundations have been laid. Koko love!

KOKO'S ACCOMPLISHMENTS

Koko has done for gorillas what Washoe has done for chimpanzees and Kanzi for bonobos; she has demonstrated that her species is capable of a level of comprehension and production of language that was not thought possible. Most laypeople in particular pictured gorillas as large, powerful, fearsome, and somewhat stupid, certainly not as bright as the frolicsome chimpanzee. Koko has disabused us of that misconception. Gorillas were not supposed to be smart enough to recognize themselves in mirrors. What then are we to make of the famous picture of Koko applying lipstick to her lips with the aid of a mirror? That awesome demonstration of self-awareness may have done more to humanize gorillas than a host of scientific tomes. However, Dr Patterson has also performed a version of the formal mirror test of self-recognition by Koko.[2] Clown paint was surreptitiously applied to Koko's brow, and her behavior in front of a mirror was compared to her behavior in front of the same mirror before paint was applied. The striking result was that Koko touched her forehead 82 times after she had seen the paint in the mirror, compared to one time per session when the paint was not present. That, together with a host of other informal observations of Koko grooming her underarms in the mirror, admiring the effects of a wig, and repeatedly touching a small spot of black pigment on her gum that was invisible without the mirror, is enough to convince all but the most recalcitrant skeptics that gorillas can be self-aware.

Koko was born on July 4, 1971, which is probably why Hanabi-Ko, Japanese for "fireworks child," was the winner in a contest to name the infant. Her nickname became Koko, and her career has been more than exciting enough to justify the "fireworks child." Penny and Ron started working with Koko when she was a little over a year old, on July 12, 1972. It was not possible for them to adopt the Gardners' sign-only technique in teaching sign language to Koko because her initial training was in the very public setting of the children's zoo in San Francisco. Thus Koko was exposed simultaneously to signing and to vocal translations of the signed message. However, all of Koko's formal training was in sign; if she were to understand English, it would be up to her to learn by observing, just as human children do.

That Koko did quite well at observation learning was a surprise. Soon after she was $2\frac{1}{2}$ years old, she signed CANDY when she heard a visitor say it.[3] It was not long after that before Penny and Ron had to resort to spelling when they did not want Koko to know what they were saying, just as Sue Savage-Rumbaugh later had to do with Kanzi.

The Koko project was part of Dr Patterson's doctoral work at Stanford University, and it has consumed her postdoctoral life as well! Koko's rate of acquiring signs paralleled Washoe's. Despite the less than ideal setting, she had imitated the signs for "food" and "drink" within 2 weeks,[3] and she learned about one sign a month, so that after 18 months of training she knew 22 signs; Washoe had 21 signs after 18 months. When she was 3 years and 3 months old, Koko had met Penny's criterion of acceptance for 78 signs. Thus her rate of acquisition increased greatly after the initial period of getting the idea of signing. After 51 months, Koko had 161 signs, while Washoe had 132 at the same age. However, Penny gave Koko credit for eight signs designated by simply pointing to body parts like the ear and nose, and the Gardners classified all of these signs as THERE. Making that correction reduces the difference to 154 versus 132. Even that difference should not be taken seriously; Penny had the advantage of knowing what worked well for the Gardners, only one animal of each species was compared, and Penny's simultaneous use of English and sign was a radical departure from the Gardners' sign-only procedure. We have no way of knowing whether that made a difference in Koko's rate of acquisition.

Koko signs to Penny that she would like a sip of her drink. It is abundantly clear from this photograph that gorillas are not quite the fearsome creatures that they used to be in popular imagination. Photo courtesy of Ronald Cohn.

We do know that Koko attempted to imitate unvoiced sounds, probably as a consequence of her training method. Had she been trained sign-only, it is unlikely that she would have made these attempts. Koko was about 30 months old when Penny noticed her first attempt to imitate. That was unexpected; Viki's babbling, as you may recall, started and stopped before she was 6 months old!

The rate of Koko's acquisition of signs does not allow any conclusion about whether gorillas are slower, faster, or the same as chimpanzees in acquiring vocabulary. That is amazing, considering that the consensus of pre-Koko opinion was that gorillas were too aggressive, too stubborn, or just too stupid to acquire any linguistic ability. The facts now force us to the conclusion that gorillas are approximately as able as chimpanzees, give or take a few percentage points in favor of one or the other. The above comparisons were based on the Gardners' criterion for concluding that an ape had acquired a sign—usage of the sign on 14 consecutive days. Penny also used less stringent criteria for other vocabulary counts: the sign had to be noted by two independent observers, and had to be used "spontaneously and appropriately" on at least half the days of a month. Using these two criteria, Koko's vocabulary grew to 200 signs by the time she was a little over 5 years old. According to an even less rigorous criterion, that she had used a sign "spontaneously and unprompted," she had a vocabulary of about 600 signs at that time, and Penny claims that Koko eventually achieved a vocabulary of about 1,000 signs. That is, of course, a quite lax criterion; Koko may well have forgotten many of those 1,000 signs, although when a situation calls for it she often comes up with a rarely used sign like QUEEN, GOSSIP, or BIG TROUBLE.

Controlled testing of Koko's comprehension indicated that her comprehension of sign was almost precisely matched by her comprehension of English. Her comprehension increased slightly when messages were delivered simultaneously in sign and vocal language. Koko's ability to understand English was surprising, considering that one theory of speech perception had as a key component the assumption that speech could only be comprehended by matching what was heard to a representation of the motor movements required to produce the speech. Because Koko was incapable of producing vocal language, that could not be the process via which she understood English. (That also applies to the bonobo, Kanzi, who also understands English.)

Penny tested Koko's comprehension by administering the Assessment of Children's Language Comprehension.[4] Koko's answers ranged from a low of 30 percent correct to a high of 72 percent correct; chance levels of responding range from 20 to 25 percent. Koko's accuracy was below that of both normal and handicapped human children, but was generally well above the chance level.

Despite the many difficulties attending giving an intelligence test designed for humans to a gorilla, Penny also administered the Cattell Infant Intelligence scale and later the Peabody Picture Vocabulary tests to Koko. Although the IQ numbers are not very meaningful given the different rates of maturation of gorillas and human children, Koko's measured Intelligence Quotient was consistently between 70 and 90. More significant is the fact that Koko's estimated mental age increased from 10.8 months when she was about 14 months old to 4 years 8 months when she was $5\frac{1}{2}$ years old. Apes' mental ages are far outstripped by humans' as both grow older, but Koko's performance is nevertheless surprising and impressive.

"Conversations with Koko" occurred daily, and led to a regular feature in *Gorilla, the Journal of the Gorilla Foundation*. Part of a conversation taken from a random selection from the journal, which happened to be the June, 1991 issue, is as follows:[5]

Koko: ELBOW TICKLE BAG. LOVE BAG. HURRY DO. VISIT BAG HURRY THERE. DO.
Penny: Can you say Wendy?
Koko: DON'T KNOW. RON THERE. (Ron has some nuts.) NUT GIMME. Koko eats the nuts Ron gives her.
Koko: NUT. BAG-VISIT HURRY THERE. LOOK DO THERE. ABOVE. (for hat?) THERE. GLASSES THERE. (For sunglasses from the bag.) HAT. (For a red sun visor from the bag.) (p. 5)

That is less than a third of that conversation with Koko. She may have been particularly loquacious on that occasion because it was her first meeting with Wendy, who had brought her some presents. Because it was her first meeting with Wendy, she of course did not know Wendy's name. Such conversations are, as I said above, a common—probably daily—occurrence, and literally hundreds have been reported in *Gorilla*.

In another conversation[6] Koko demonstrated her ability to engage in fantasy play. On April 25, 1987, Penny recorded the following observations:

Penny observes as Koko takes a doll and a multicolored pencil to her back room. She uses the pencil as if applying lipstick to her top and bottom lips and to write on an envelope. Koko takes her baby doll outside to her chute leading to the playyard. Koko: BABY (over her doll). K: COME (To Penny, for her to follow Koko to the back room) K: LOVE (To her doll, which she then strokes and kisses). She puts the baby into the purse it came in and zips it closed, puts the strap over her arm, and signs K: NO-THERE (pointing to the purse). COME. COME TICKLE (To Penny, who declines this invitation to follow her to the back room.) Koko settles in the back room with the baby doll and her gorilla Koko doll. DRINK NIPPLE, BABY (Cradling her baby doll). K: NIPPLES (two hands) DRINK THERE (to gorilla doll's mouth). Koko signs as she holds the baby doll in her lap. K: YOU BOY (Or YOU MIKE. The signs are similar, and Koko signed away from Penny). P: Careful with her head or his head (Voice only). Koko was pulling the neck of the baby doll. K: GORILLA HAVE (over the doll held on her chest). FOOT (Koko's sign for *male*, signed on the baby doll's foot). Then Koko mouths the doll's foot. K: YOU (indicating the baby doll). Then she cradles the baby at her stomach. K: THAT (on the gorilla doll's bellybutton area). Then Koko puts the baby doll

on the gorilla doll's stomach, and signs. K: STOMACH (on the gorilla doll's
stomach) (p. 8).

Penny commented on this conversation as follows: "Although Cathy Hayes
remembers her imaginary play with the chimpanzee Vicki as 'a symbol of…
the tragedy of the language barrier that separates us' our understanding of
Koko's solitary games is reinforced and deepened by the bridge of her
language, allowing her to clarify her actions" (p. 8).

Koko thus illustrates one of the highly significant reasons for the study of
animal language, that it provides a window, more transparent than the one
afforded by non-linguistic behavior, on the feelings and thought processes of
our fellow travelers on planet Earth.

The simultaneous use of English and sign also enabled Koko to rhyme,
indicating at the least that she recognized similarities in sound. Barbara Hiller,
a long-time worker with Koko, reported the following conversation with Koko
about a series of toy animals arranged in front of them:[4]

Barbara: Which animal rhymes with hat?
Koko: CAT.
Barbara: Which rhymes with big?
Koko: PIG THERE. (She points to the pig.)
Barbara: Which rhymes with hair?
Koko: THAT. (She points to the bear.)
Barbara: What is that?
Koko: PIG CAT.
Barbara: Oh, come on.
Koko: BEAR HAIR.
Barbara: Good girl. Which rhymes with goose?
Koko: THINK THAT. (Points to the moose.) (p. 141)

In addition to her rhyming ability, Koko appears to have an innovative sense
of humor; for example, she pretended that her nose was thirsty, made the sign
for drink (which she had used hundreds of times) on her ear rather than her
mouth, and said that she was an ELEPHANT GORILLA after using a tube that,
with a little imagination, might resemble an elephant's trunk. Koko also used
her imagination in designing insults, for example, in sometimes calling
Michael, her male gorilla companion, a DIRTY TOILET or TROUBLE DEVIL or
some combination of these insults. Other words of opprobrium applied to her

caretakers when they failed or refused to get food or drink quickly enough included ROTTEN, BIRD, STUPID, STINK, and STUBBORN-DONKEY.

Koko also invented unique combinations of signs, for example, EYE HAT for a mask, BOTTLE MATCH for a lighter, ELEPHANT BABY for a Pinocchio doll, and WHITE TIGER for a zebra. In this respect, Koko's creativity is similar to that of other enculturated and language-trained animals like Washoe and Chantek, the orangutan. Finally, she sometimes uses language to further her humorous tricks; for example, she once tied Penny's shoelaces together and then signed CHASE.[7]

KOKO AND MICHAEL

On September 9, 1976, when Penny's project was 4 years old, Koko was over 5, and he was about $3\frac{1}{2}$ years old, a male gorilla joined the project. When he arrived his name was King Kong, but Penny quickly changed it to Michael. The hope was that Koko and Michael would produce a baby. Penny was well aware that animals raised together would probably not breed, so she limited their playtime together. Koko was older and larger, so she dominated Michael at first. Perhaps for that reason, or perhaps because they were too much together, Koko never showed a sexual interest in Michael. Michael was interested, but Koko never solicited him. Michael learned many signs, but as he matured he became less communicative than Koko. In the editorial of the memorial issue for Michael after he died on April 19, 2000, Penny said that Michael eventually learned about 500 signs to Koko's 1,000 (she does not say what criterion she was using to arrive at these figures, but it evidently was the relaxed "spontaneous and appropriate use" criterion). Michael achieved a considerable reputation as an artist, and his works were exhibited in San Francisco. Like Koko, he showed clear signs of recognizing himself in a mirror, thus replicating the finding that gorillas are capable of self-recognition.

Koko and Michael were later joined by Ndume, who was even more dominated by Michael than Michael had been initially by Koko. Although Koko had chosen Ndume from among videotapes of several gorillas, she and Ndume have shown little or no sexual interest in each other.

THE GORILLA FOUNDATION

Penny believes that the limitations of space and climate in Woodside are not conducive to breeding by gorillas, and she is continuing a longstanding and intense effort to obtain enough backing to move Koko and Ndume to a warmer climate. She has supported herself and her gorillas since 1976 through the Gorilla Foundation, Box 620530, Woodside, CA, 94062-0530 (Internet address www.gorilla.org, or www.koko.org). Recently 70 acres on Maui were donated to the Gorilla Foundation, and work on a facility has begun, despite the fact that not enough money is available to complete the project. Time is of the essence because Koko has reached the middle age (for gorillas) of 30+ years, and will not remain a potential mother for many more years. The latest news on the Maui move and on the activities of the Foundation is available on the web site. As of this writing, the Foundation has 40,000 members, all of whom receive *Gorilla*, published twice yearly with information on both the gorillas' and the humans' activities.

KOKO'S CONTRIBUTION

It is not really just Koko's contribution—Michael and Ndume also count, and certainly there would have been no contribution without the heroic efforts of Drs. Patterson and Cohn. More accurately, we should say "the contribution through Koko." Whatever we call it, the contribution has been enormous. Koko's accomplishments have forever changed humanity's view of the gorilla and, through that change, our view of all nonhuman animals. We no longer think of gorillas as fearsome and aggressive beasts, nor as less intelligent than chimpanzees; Koko's progress in producing and comprehending sign closely paralleled Washoe's, and her comprehension of English is stunning.

If our world ever becomes kinder and gentler (as it must if a reasonable world is to exist), history will record that the gorilla Koko was one of humanity's great benefactors, a kind of animal equivalent of Mahatma Gandhi or Nelson Mandela. It is enough. Koko fine gorilla.

REFERENCES

1. S. Wise, *Rattling the Cage: Toward Legal Rights for Animals* (Perseus Publishing, Cambridge, MA, 2000).
2. F. G. Patterson, Self-awareness in the gorilla Koko, *Gorilla: Journal of the Gorilla Foundation*, 14(2), 2 (1991).
3. F. G. Patterson and R. N. Cohn, Language acquisition by a lowland gorilla: Koko's first ten years of vocabulary development. *Word*, 41(2), 97–142 (1990).

4. F. G. Patterson and E. Linden, *The education of Koko* (Holt, Rinehart, Winston, New York, 1981).

5. F. G. Patterson, Conversations with Koko. *Gorilla: Journal of the Gorilla Foundation*, **14**(2), 5 (1991).

6. F. G. Patterson and M. M. Kennedy, Fantasy play. *Gorilla: Journal of the Gorilla Foundation*, **20**(2), 8 (1997).

7. J. Gamble, Humor in apes. *Humor*, **14**(2), 163–179 (2001).

CHAPTER EIGHT

Chimpanzees can Write with Plastic Symbols

BYPASSING THE VOCAL CHANNEL

The Premacks, Ann James and David, were interested in animal cognition and language very early, at least as early as 1954 (personal communication, David Premack, 1954) and started preparatory work with monkeys soon after that time. In 1964,[1] the Premacks adopted two young female chimpanzees named Sarah and Gussie. Sarah proved to be an excellent student, but Gussie never learned a single word. The Premacks later worked on language training with several other animals, but none of them were as intelligent as Sarah. A characteristic of the Premacks' work is that their primary interest was in the cognition of chimpanzees, with language regarded more as a window to the chimpanzee mind than as the center of their attention. David Premack's discussion[2] of the relative problem-solving abilities of language-trained and non-language-trained chimpanzees makes that clear. He found striking individual differences in intelligence between chimpanzees in each group, whether language-trained or not: "We have ... had both gifted and ungifted animals in each group. Sarah is a bright animal by any standards, but so is Jessie, one of the non-language-trained animals. The groups are also comparable at the other end of the continuum, Peony's negative gifts being well matched by those of Luvy" (p. 125).

At nearly the same time that Washoe arrived in Reno to start her project with the Gardners, David Premack (with Schwartz) described his first method for circumventing the vocal channel. He planned, and had constructed, a sound-generating apparatus that produced different phonemes when a joystick was moved to different locations. The plan was to teach a chimpanzee to "talk" by manipulating the joystick.

However, constructing words from phonemes, of which English uses about 45, is probably at least as difficult conceptually as spelling words with letters, of which English uses only 26. All of the successful animal language programs use units that are glossed as words. The joystick idea was ingenious, but was too complex to work with chimpanzees.

Ann and David Premack then developed a second, much more successful method of circumventing the vocal channel. Like the first method, it contrasted markedly with the Gardners' use of simplified American Sign Language. Ann, who was born in China, was inspired by the pictographic characteristics of Chinese writing to suggest using plastic chips of different shapes to substitute for English words. One of David's students, Jim Olsen, a surfer who was adept at repairing surfboards with plastic, furnished them with plastic "loaves" that were sliced into various shapes. The chips had a metal backing and could therefore be stuck to a magnetized board, just as you stick notes on your refrigerator. A sequence of chips was analogous to a written sentence. Thus the Premacks' chimpanzees, like those in the project by Duane Rumbaugh, were writing and reading rather than speaking and listening. This fact gave Ann Premack license to title her book, *Why chimpanzees can read*.[3]

THE PREMACKS' REARING AND TRAINING TECHNIQUES

The rearing techniques used by the Premacks were a departure from the home-and-family approach of Kohts, the Kelloggs, the Hayeses, the Temerlins, and the Gardners. Their work with Sarah[4,5] was more similar to the general approach and methods used by experimental psychologists in animal laboratory research, and in some ways resembled the combination of rigor and socialization employed later by Duane Rumbaugh with Lana and by Terrace with Nim. After her first year in a home setting, Sarah was moved to a laboratory. Nevertheless, she continued to have close social contacts with humans and to have daily outings, which included romps on the beach, until she was 7 years old (the Premacks had moved from the beachless University of Missouri at Columbia to the near-the-beach University of California, Santa Barbara).

Sarah's language training was limited to specific hours each day, typically four, in a controlled laboratory setting. Sarah and the experimenters could not conveniently carry the plastic symbols and the magnetized board on which they were placed as they went on outings or moved about within the laboratory. The methods used in Sarah's training sessions were precise. The Premacks'

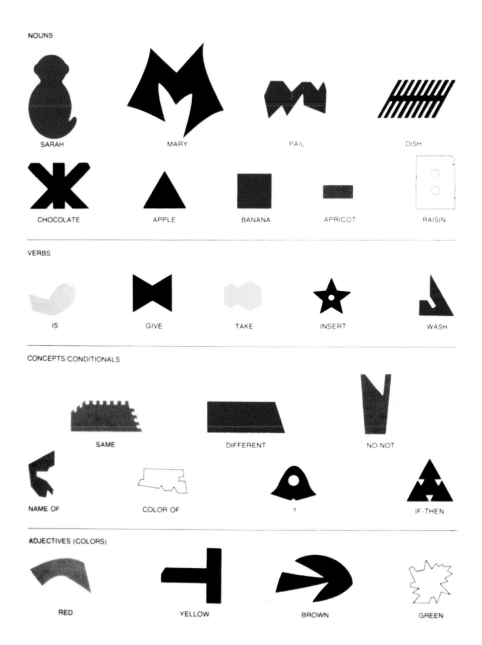

NOUNS

SARAH • MARY • PAIL • DISH

CHOCOLATE • APPLE • BANANA • APRICOT • RAISIN

VERBS

IS • GIVE • TAKE • INSERT • WASH

CONCEPTS/CONDITIONALS

SAME • DIFFERENT • NO-NOT

NAME OF • COLOR OF • ? • IF-THEN

ADJECTIVES (COLORS)

RED • YELLOW • BROWN • GREEN

This photograph is a sample of the different classes of plastic symbols that the Premacks used with Sarah and the other chimpanzees. Each symbol had a metallic back that adhered to the magnetic board. Photo courtesy of Ann and David Premack.

philosophy differed from that of most language researchers in that they believed that the scientific value of an animal was reduced when it was studied in uncontrolled interactions with humans or other animals. Thus each of Sarah's tasks was designed to inculcate a specified linguistic function and to evaluate how well Sarah learned it. The Premacks were interested in the chimpanzee's intelligence more than they were interested in how well a chimpanzee could simulate human language. They were undertaking a functional analysis of language skills.

SARAH'S TEACHING AND WHAT SHE LEARNED

For the Premacks, language training was a way of (1) mapping existing knowledge through the acquisition of word meanings, (2) helping the chimpanzee to master a set of concepts like name-of, color-of, same as, different from, etc., and (3) teaching rules for relating words to one another in order to construct and decode sentences, for which word order is relevant.[6] As important as vocabulary is, it is not the only aspect of language. The Premacks never pressed their subjects to learn a large vocabulary; rather, they stressed the acquisition and use of concepts, together with the production and understanding of syntactical relations. Nevertheless, they estimated Sarah's eventual vocabulary at about 130 words, about the same number as Washoe mastered during the first few years of her training. They do not describe the criterion or criteria by which they determined her vocabulary size, but I presume that she responded correctly in tests using 130 plastic shapes.

Sarah's language chips had different colors and different random shapes that bore no obvious relationship to their referents. When they were placed on the magnetic board, they allowed continued access to what had been written, and made it unnecessary for Sarah to keep track of what she had already written. Sarah, and later the other chimpanzees trained by the Premacks, did not like to write in the traditional English left-to-right word order, so they were allowed to use the top-to-bottom arrangement that they preferred. Sarah had access to a limited number of plastic shapes at any one time, so her linguistic tasks resembled multiple-choice tests. Under these conditions, it was much easier to see and record Sarah's responses than it was to determine what Washoe did; Washoe could make any available sign or any non-sign movement at any time, and it was up to the observer to decode what she had done or signed. Pinker, among others, was extremely critical of this aspect of the Gardners' work.

The Premacks assumed that every complex linguistic behavior could be decomposed into simpler units. The basic strategy was to discover and teach the units so that language could be inculcated in animals that otherwise did not have it. Their chimpanzees taught the Premacks that there were even more steps than they would have expected.

They first tried to teach Sarah by having her observe social interchanges and pairing the plastic elements with aspects of the interactions. One involved a baby bottle filled with milk, of which Sarah was still fond when training began. Although this technique resembles the one used later successfully by Irene Pepperberg with her parrot Alex, it did not succeed with Sarah. The mixed modalities of spoken English and plastic words may have been too confusing for Sarah, and the situation could have been too complex; whatever the reason, the approach was abandoned.

Sarah's first success in learning words resulted from pairing objects with the corresponding plastic language elements placed on a language board. Thus Sarah and later the other chimpanzees were taught to place the symbol for apple on the board to obtain a helping of apple. Only the APPLE symbol was available. After this errorless training (she had no other symbol with which she could make an error), she was given two symbols, APPLE and BANANA and required to put the correct one on the board in order to obtain her reward. It required near-heroic patience to get the chimpanzees to recognize that different words went with different fruits. However, they eventually did learn to put the correct piece on the board, although there may have been a tendency to continue to choose the word representing the favored fruit, rather than the one that was available.

Once progress had been made on distinguishing objects, the verb GIVE was introduced, and Sarah was required to place it on the board along with the name of the fruit. Again, no error was allowed in the verb choice; only GIVE was available. As with the fruit names, a second verb, INSERT, was added once Sarah had learned to construct two-word sentences. Names of persons, words for constructs like same and different, and adjectives were added later in the step-by-step procedure.

The Premacks have from the beginning been concerned with the cognitive processes that underlie language acquisition. For example,[7] they analyzed the progression of processes in their chimpanzees as they learned to use different plastic tokens to request different fruits. The Premacks believed that the chimpanzees entered the experiment with the class properties of fruit already in place. They then learned that symbols, and only symbols, were to be placed on the magnetic board, and later the mistaken rule that any symbol could be exchanged for any fruit. They never placed anything but a magnetized symbol on the board, nor did they accept anything but fruit in exchange for the symbol. (Although they liked chocolate better than fruit, they took fruit rather than chocolate when both were offered after a symbol was placed on the board.) Only when the properties of the two classes of things, symbols and fruit, were in place did they unlearn the "any symbol gets me any fruit" rule and form the correct connections between the specific symbol and the fruit that it named.

After Sarah had learned the names of several concrete objects, she was taught NO as an injunction against certain actions, as in NO SARAH TAKE GRAPE. Sarah was taught the concept of "same" by presenting her with pairs of identical objects and requiring her to place the symbol SAME between them. DIFFERENT was the symbol that was required for non-identical pairs. Four different types of wh- questions were used in further training. For example, (1) which of two alternatives is the same as the spoon (a real spoon is the sample, and a spoon and clothespin are the alternatives)? (2) Which of the two alternatives is not the same as the sample? (3) What is the relationship of the spoon to the spoon (SAME, DIFFERENT)? What is the relationship of the spoon to the clothespin (SAME, DIFFERENT)?

After she knew these symbols and could answer the questions above for different sets of objects, she could be given an identical pair with SAME between them, followed by the interrogative symbol, and required to place YES as the answer. For example, a real banana, SAME, another real banana, QUESTION, requires a YES because they are, indeed, the same. Thus non-identical pairs with SAME between them required a NO answer. Identical pairs with DIFFERENT between them required a NO answer, and, finally, non-identical pairs with DIFFERENT between them required a YES answer.

The "interrogative particle" placed after a string or between objects changed a declarative statement into a question. For example, SPOON? SPOON asked Sarah to indicate whether the two spoons were the same or different; SPOON DIFFERENT SPOON? asked Sarah to indicate whether the statement was correct or not. Sarah was supposed to judge the truth of the statement and simply answer YES or NO depending on whether a statement, say RED ON YELLOW corresponded to the true state of affairs. But Sarah seemed to prefer truth to fiction, so she often took away the green that was actually there, replaced it with the yellow so that the statement was true, and answered YES. As David Premack put it, "That already suggests Sarah's use of language was not humanlike, not entirely truth-judgmental. She liked correspondence or equivalence and brought it about when she could (personal communication, David Premack, 2002)."

Once Sarah had learned same and different, she was also taught the concept of "similar." For example, when she was asked to judge the relationship between a larger and a smaller red square, she judged them to be similar rather than same or different, and she judged other objects that were the same on two of three dimensions in the same way.

Sarah also mastered more complex judgments of sameness versus difference. Pairs of apples were judged to have the same relationship as pairs of bananas, but to be different from an apple and a banana. This required her to judge the relationship between relationships, a higher order of conceptualization than judging relationships between objects.

The Premacks taught Sarah the concept NAME-OF, and once she got that idea it was much easier to teach her the association between a symbol and its referent. She also understood COLOR-OF and after she was told BROWN COLOR-OF CHOCOLATE she was able to select the brown disc from a set of four distinctively colored discs, with no additional training on what BROWN meant. That behavior was a remarkable selective association between one characteristic of chocolate and the physically dissimilar symbol for brown.

Sarah also learned to describe whether a card of a given color was on, under or to the side of a card of another designated color, and she learned to place them correctly in response to requests like RED ON GREEN. Sentences were built up through the acquisition of unit skills that could be hierarchically organized into structured strings of words. Subsequently, mastery of the conditional IF-THEN allowed Sarah to respond appropriately to statements like [IF] SARAH NO TAKE RED [THEN] MARY GIVE SARAH CRACKER (the bracketed words were actually a single if-then symbol).

The Premacks have always been concerned about the fundamentals of language. One such fundamental question is "When does something (a plastic shape, e.g.) come to function as a word?" One such test was passed when Sarah showed that she regarded APPLE as having the properties of the object for which it stood. She described the blue piece of plastic as being red and as having a stem. No apple was present to elicit these responses, and the plastic itself had neither of the properties Sarah ascribed to it.

THE EFFECTS OF LANGUAGE TRAINING
ON COGNITION

The Premacks demonstrated[1] some remarkable changes in chimpanzees' cognitive abilities as a function of language training. A striking example was the language-trained animals' ability to complete sequences like the following: APPLE-BLANK-CUT APPLE. The language-trained animals all succeeded in substituting a knife into the blank when presented with a real apple, a space for another object, a cut piece of apple, and alternatives like a knife, a pencil and a sponge that could be placed in the blank space. None of the non-language-trained animals succeeded at this type of task. Their most frequent substitution into the blank space between an apple and a cut apple was the pencil, presumably because it happened to be red. They seemed to be incapable of learning that what was wanted was an action that connected the two objects, not something that shared some physical characteristic.

Although the Premacks could not provide a formal description of the tasks on which the groups of chimpanzees differed, there was a tendency for the tasks on which only the language-trained animals succeeded to be more abstract than those that both groups could perform successfully. They suggested that non-language-trained animals were less able than language-trained animals to use an abstract code for solving problems. They appeared to lack a second, representational system that allowed the language-trained animals to solve the problems.

The Premacks concluded[8] that it was not language training in general, but specifically learning to use SAME and DIFFERENT appropriately that accounted for most or all of the apparent differences between language-trained and non-language-trained animals. Language training did very little unless it included the distinction between same and different. Only with "heroic" training were animals without words for same and different able to perceive relationships between relationships, and when they did perceive one such relationship it did not transfer to other cases. For example, after learning SAME and DIFFERENT chimpanzees could judge AA as SAME as BB because the relationships between both pairs was SAME. In addition, apes that knew SAME and DIFFERENT transferred to new cases of relationships between relationships with little difficulty. They could judge things like half an apple-whole apple as having the SAME relationship as half a glass of water-full glass of water.

An alternative possibility that has been suggested to account for the differences between language-trained and non-language trained chimpanzees is that language-trained chimpanzees are better at using images to solve problems. However, it is difficult to explain how AA can be judged to have the same relationship as BB by superimposing images; on the other hand, if both AA and BB are labeled SAME it is easy to label the relationship between relationships as SAME.

The experiments described above illustrate the two aspects of language training that were probably of primary interest to the Premacks: that language training allowed animals to reveal their cognitive processes more easily, and that the processes were themselves changed by the training. They suggest that humans may share an abstract code only with primates, whereas our "imaginal code" is held in common with many other animals.

Sarah Boysen[9] demonstrated a remarkable quality of this abstract code. She first taught her chimpanzee Sheba to associate small numbers of objects with their corresponding Arabic numbers. She then had Sheba play the following game: two trays holding different quantities of desirable food were shown to Sheba. Whichever tray Sheba chose went to another chimpanzee, and Sheba got the remainder. Sheba was presumed to want the larger portion, but she was utterly unable to choose the smaller portion for her partner. However, when Arabic numbers designating the size of the food cache on the two trays were substituted, Sheba immediately chose the smaller portion for her partner. It appears that there is much less emotion attached to the abstract code than to the corresponding concrete objects. Perhaps much of our human success is related to this feature of language!

David Premack's beliefs about the importance of grammar shifted over the years, as indicated by the following quotation (personal communication, David Premack, 2002):

> ...my approach was functional or psychological not grammatical; I conceded that grammar was unique to humans (that is, human grammar with all its special complexity), but argued that grammar of human complexity was not required to realize those functions that cannot be realized without language, or realized nearly as well. I was a little naive...holding that only rules which underlie mental functioning are of interest. Today I would grant the intriguing special role grammar plays in realization of language-unique functions.

Thus David Premack's position (presumably Ann's is similar) is that a complex grammar is an exclusively human possession, which he believed all along. However, he now believes that this human-unique grammar is of greater adaptive importance than he thought at the beginning. Probably no other animal language researcher has thought more about this problem than David Premack, so his opinion, although it is a little discouraging with respect to the limits of animal competence, must be taken seriously.

THE ANIMAL MIND

Finally, David Premack asks[10] whether animal and human minds are better represented by overlapping circles or by concentric circles with animal abilities totally contained within the larger circle of human intelligence. If the circles only overlap, then human and animal intelligences have some things in common, but each has unique elements not shared by the other. If animal intelligence is wholly encompassed by human intelligence, then it is assumed that human intelligence is wholly superior.

Mark Hauser in his book *Wild Minds*[11] deals with a similar issue. He uses the concept of "mental toolkits" as the centerpiece of his analysis of intelligence. Animals, says Hauser, share with humans several basic toolkits, for example, a set of principles for recognizing objects and predicting their behavior. Hauser also credits all animals with special toolkits for processing information about objects, numbers and space. Humans, of course, are much better than animals at processing numbers, and also have toolkits for processing objects and spatial relationships.

So far then, we have seen only cases in which human intelligence is equal to or greater than animal intelligence. Perhaps the wild mind is altogether inferior to the human mind. Clearly, as Hauser and Premack would agree, humans are vastly superior to all earthly animals in their ability to learn and use language. However, if we accept this view we have jumped to the wrong conclusion. Animals do have unique abilities not shared by humans. Perhaps the most convincing case is dolphins' talent for acquiring and processing acoustic information. This skill is highly adaptive and is a type of intelligence according to any reasonable general definition. Furthermore, not only can unaided humans not match dolphins' ability, but also the best human-designed transducers, computers, and software cannot match their ability. Only by adopting radically speciesist definitions of intelligence could we deny that this ability is a type of intelligence. There are many similar examples of specialized intelligence in animals, including the ability of migratory birds to navigate, of dogs to analyze odors, and even of bees to navigate using polarized light. It may be difficult for humans to accept the idea that a bee brain can process information that we cannot even sense!

Thus clearly the partially overlapping circles proposed by Premack are the correct model for representing the relationship between animals' and human intelligence. Research on language in animals is motivated in part by an attempt to increase the overlap in the circles of animal and human intelligence.

THE IMPACT OF THE PREMACKS' RESEARCH

The Premacks started their animal language research even earlier than the Gardners, although their first publications on work with plastic symbols were later than the Gardners' first publication on chimpanzee signing. Their work was influential in part because of their care in defining the properties of language as they saw them. Their analyses and the systematic training procedures to which they led were also impressive to other researchers. Throughout it all, and despite their careful behavioristic approach to experimentation, they have been willing to speak mentalistically and acknowledge nativistic factors, as when David Premack says,[6] "The mind appears to be a device for forming internal representations. If it were not, there could be neither words nor language … every response is a potential word. The procedures that train animals will also produce words" (p. 226).

The Premacks conducted systematic observations on whether chimpanzees could produce words that they comprehended or comprehend what they had learned to produce. The results are complex and not always clear, but it appears that transfer from one mode to the other starts out very poor but becomes increasingly effective as the chimpanzee becomes more accomplished with language. In the final stages of her training, Sarah was immediately able to use words that she had learned to comprehend, and to comprehend words that she had learned to use. In addition, once she had learned the meaning of NAME OF she could ask for the name of an object and very quickly attach a new plastic word to the object.

Like Terrace, the Premacks have concluded that chimpanzees do not construct true sentences. For them, a preference for producing correct word orders, or using word order to determine responses, is insufficient evidence of an ability to construct sentences. Sarah could, as Ann's book title suggested, read, but what she wrote were not true sentences. The Premacks did not, however, deny that language-trained chimpanzees possessed some language. On the other hand, they were impressed by the vast gulf between the language abilities of chimpanzees and those of even very young human children. No doubt their work and point of view have, or should have, made other researchers more careful about claiming syntactic abilities for their animals.

Thus the Premacks' experiments show that chimpanzees and children differ greatly in the ability to perceive and use syntactic patterns that encode conceptual relations. However, both species can see the conceptual relations themselves. In this respect we see the Premacks using language as a window to the mind, as Margaret Floy Washburn suggested we should do in the quote in the first chapter of this book. Finally, their research provides a model for the systematic and progressive training of symbolic behaviors and for how the processes underlying these behaviors can be identified.

REFERENCES

1. D. Premack and A. J. Premack. *The mind of an ape* (Norton, NY, 1983).
2. D. Premack, The codes of man and beasts. *Behavioral and Brain Sciences*, **6**(1), 125–137 (1983).
3. A. Premack, *Why chimpanzees can read* (Harper and Row, NY, 1976).
4. D. Premack, The education of Sarah. A chimpanzee learns the language. *Psychology Today*, **4**(4), 54–58 (1970).
   D. Premack, Language in chimpanzee? *Science*, **172**(3985), 808–822 (1971).
5. A. J. Premack and D. Premack, Teaching language to an ape. *Scientific American*, **227**(4), 92–99 (1972).
6. D. Premack, *Intelligence in ape and man* (Erlbaum, Hillsdale, NJ, 1976).
7. D. Premack and A. Premack, Waiting for manifesto 2. *Behavioral and brain sciences*, **23**, 784–785 (2000).
8. L. Weiskrantz (Ed.), *Thought without language. A Fyssen Foundation symposium.* (Clarendon Press, NY, 1988) pp. 46–65.
9. S. Boysen and G. G. Berntson, Responses to quantity: Perceptual vs cognitive mechanisms in chimpanzees (*Pan troglodytes*). *Journal of Comparative Psychology*, **21**, 82–86 (1995).
10. D. Premack, Minds with and without language, in: *Thought without language*, edited by L. Weiskrantz (Clarendon Press, Oxford, 1988), pp. 46–65.
11. M. Hauser, *Wild minds* (Henry Holt, NY, 2000).

CHAPTER NINE

Lana Learns Lexigrams

In this chapter Duane Rumbaugh speaks in the first person about his 30+ years of research on teaching animals, in particular the chimpanzee Lana, to use lexigrams.

LESSONS ABOUT LANGUAGE FROM STUDIES OF THE APE

Since language research with apes was revitalized, notably by Gardner and Gardner, a crescendo of interest and criticism has made it difficult to maintain a balanced and constructive perspective about the value of the research. Researchers initially presented an array of justifications for their efforts, ranging from performing an experimental analysis of language to opening up free-flowing visits with the hope of understanding apes' dreams and innermost perspectives on life and the world.

The focus of the present chapter is on the chimpanzee Lana (*Pan troglodytes*), and how she has helped to alter people's view, not only of herself and her species, but also of language. In an editorial[1] in *The New York Times* the writer asked, "Is she or is she not 'talking,' that is, using language similar in some degree to human use of language? ... In short, Lana's virtuosity forces us to ask where the boundary line is, or alternatively how large and flexible her command of language need be, before even the most skeptical agreed that she had broken the verbal barrier." Lana "talked," and she was heard in the far reaches of the nation and the world. Who was Lana? What was work with her about?

Lana already had her name when she joined the language project I conceived in the winter of 1970. Perhaps she was named after Lana Turner, the beautiful actress of yesteryear. I do not know for sure, but I do know that her name provided me with a handy acronym, LANA: the Language ANAlogue Project. Why analogue? Because when I started the ape-language project I did not believe for one second that apes were capable of true language. Sure, they could dupe us and behave in ways that made us want to believe that they understood what we said, but they simply could never acquire language. I was an avid behaviorist who did not believe that any life form other than us (arrogant) humans could or would think, let alone acquire language. I had been taught in graduate school that, because humans have language, we can think and do related things with language. Because animals did not have language, they just could not think! It was as simple as that! Now, however, I look back upon that time and my graduate school studies of the early 1950s and shake my head. How arrogant! How dogmatic! How self-serving! How prejudiced we were to draw the line with us on one side, the thinking side, and animals on the other side, the unthinking side. We felt comfortable in that conclusion, that dictum, and there we would stay!

However, we *did not* stay with that view because some bold researchers, even in the first half of the 20th century, had asked in their research, "Can apes acquire language?" They had little or no success, as we have discussed earlier in this book, but they had looked in the box. Furness concluded,[2] "I regret that I am forced to admit, after my several years observation of the anthropoid apes, that I can produce no evidence that might disturb the tranquil sleep of the reverend gentleman" (p. 290). Nevertheless, the box just would not stay shut! Thank God it did not.

In the mid-1960s, Beatrix and Allen Gardner of the University of Nevada asked, "Might apes acquire sign language?" Their chimpanzee Washoe surprised everyone. She learned dozens of signs and convinced the people around her that she knew both what she was signing and what others were signing to her—at least on some occasions. Allen told me about the launching of the Washoe Project at a convention party. Sadly, at the time I was unimpressed. I was sure that all that Washoe learned and did with sign language could and would be accounted for quite simply as instances of operant conditioning—doing something because it had been reinforced with foods and social praise in the past. But Washoe proved me wrong. It was when she started to put signs together that my interest was piqued. For instance, when Washoe signed WATER BIRD in the presence of ducks that she saw on a pond, the implications were too strong to ignore. Aware of her own actions or not, it opened my mind to the possibility that an ape might know something about language and the creative use of language, as in this instance.

In 1969, I received a phone call from the late Dr. Geoffrey Bourne, asking whether I would be interested in applying for a research faculty and administrative position at the Yerkes Regional Primate Research Center of Emory University. At the time, I was a faculty member at San Diego State College, now a university.

I responded positively to Dr. Bourne's query. A trip to Atlanta followed shortly, and in a few months I was moving!

INCEPTION OF THE LANA PROJECT AND KEYBOARD

Dr. Bourne, then director of the Yerkes Center, had determined that I should study language in apes. "But Geoffrey! It's a waste of time. There's no reason to expect success in what I know will be a very difficult and expensive project. Apes can't think. Therefore, they can't have language. You know that!" Geoffrey followed several exchanges of that kind with still another "suggestion." "Do it!" That convinced me—I would do it, but how?

Despite the encouraging leads being reported from the Washoe Project, I was not inclined to use manual signs derived from the American Sign Language, because I thought it would be too labor-intensive and subject to controversy regarding what was said by the ape and others present. Videotaping was not yet feasible, and filming was very expensive. In addition, even with filmed records one had the demanding task of replaying the film over and over to translate events into a paper copy. However, computers were also very expensive and very limited by today's standards. Nevertheless, computers were attractive because they would automatically record all events and allow for a public record of what transpired. Objectivity in recording data is one of the hallmarks of good science. It was time that computers were introduced, if possible, into research on ape language competence.

I called my good friend, Professor Harold Warner, who was on the cutting edge of biomedical research and its attendant technology. Hal, as everyone called him, came right down from his third floor office to talk. I explained that we needed to get involved with research on language and apes, and I asked how he would advise bringing computers into the effort. His first point was that we would have to find a way for the computer to receive signals to process, record, and act upon. The idea of having the apes punch keys came to the fore promptly, but what would a key press mean to the computer? We were so thoroughly brainwashed by our experiences with keyboards on typewriters with which words are spelled and messages written that initially we thought along that same line for work with apes. Apes would spell words?

How ridiculous! How could we expect them to spell words when they had absolutely no notion of what a word was and what an alphabet meant? Spelling of words by apes was not the way to go. We had to break out of the box. We decided that each key would stand for a word! Yes! That held promise, although the number of words would be quite limited if we thought in terms of a computer terminal that was like a typewriter keyboard. Well, we could have as many keys as we wanted. We could have several hundred if we needed them. Ok, that is fine, but if the locations of word-keys were unchanging, as keys' positions remain constant on typewriters, the ape might learn only the positions of keys, nothing about the words represented on the surfaces of the keys. That was a real concern. To deal with that, we would have to engineer the keyboard so that the locations of the keys could be changed. Then the ape subject would have to attend to the pictures embossed on their surfaces and not just to the locations of keys. Great! But then, what would a word be for the ape? And how would words and sentences be visually conveyed to the ape and others present? What would they look like? Special effects generators were not feasible at that time, and we were not yet able to produce anything other than alphanumeric characters on a monitor or television screen.

Charles Bell, our electronics technician, found the solution to this problem. Words would be represented by distinctive geometric patterns, composed from a small set of stimulus elements to which color could be added if we wished. We had had experience with industrial projector units that were so small that many could be lined up in a row, as one might think of lining up the words of a sentence. These projectors had a housing for film, 12 lenses in a 3×4 matrix, and 12 small light bulbs. If we designed stimulus elements, such as a line, a square, rectangle, parallel lines, a wavy line, a triangle and so on, and had the option of illuminating simultaneously a number of light bulbs, we could compose a variety of patterns on these little projectors. So our words would be patterns, and they were named lexigrams by Ernst von Glasersfeld, our psycholinguist from the University of Georgia. Each lexigram would be a distinctive pattern. We also learned from Hal Warner that we could produce a variety of colors for backgrounds of the lexigrams by including three primary colors in each projector's film sheet. By selective activation of the lamps, and by controlling how brightly they were illuminated, we were able to get six or seven useable colors! Thus, Lana could "write" by pressing keys with lexigrams inscribed on their surfaces. Each time she pressed a key, the lexigram pattern was reproduced on one of the projection units above the keyboard. Lana and

Lana at her lexigram board, with the lexigrams on the keyboard in front of her, and the display units that echoed the key presses above. The increased illumination of the keys and the lighted display units told Lana which keys she had pressed and reduced the need for her to remember. Photo courtesy Dr. Duane Rumbaugh. Photo by Frank Kiernan.

the experimenters could then read what she had written, in the order that she had pressed the keys, by seeing what was arrayed on the projectors from left to right. Background colors for lexigrams corresponded roughly to semantic classes; for example, any violet lexigram signaled the name of a person or of the dispensing machine. Red lexigrams represented edibles.

Lexigrams were designed to have only a single meaning for ease of learning and so that the computer parser would have fewer complications in determining whether the ape's sentences were legal—that is, grammatical. The parser worked by indicating the relationships into which a given lexical item could be entered. The grammar, named Yerkish in honor of Robert M. Yerkes, worked with some 30 correlators that connected types of items. For example, the first one connected autonomous actors performing a stationary activity, as LANA DRINK. Another example is a movable actor changing places, as TIM MOVE. Yet another example is a predication, as BANANA BLACK. Lana, on her own initiative, called overly ripe bananas BANANA THAT-IS BLACK in naming tasks. She also coined the term BALL WHICH-IS ORANGE for the oranges; BANANA WHICH-IS GREEN for cucumber, and COKE WHICH-IS ORANGE for a Fanta soft drink, that is indeed orange in color.

Thus, in sum, we could write lexigrams on the projectors that in principle would allow Lana to produce sentences, just as we do with words all the time. The ape subject was to have ready access to an electronic keyboard on which he/she would write out sentences asking for various foods and requesting certain activities, including human companionship. A second keyboard would permit the experimenter/teacher to query the ape about things, to ask questions to determine or clarify what Lana wanted, or knew about various items and their attributes. At that time it was not possible to have key presses produce speech corresponding with the lexigrams. That was to come years later. That said, we had a promising concept of a system, and we went to work to obtain funding so that we could put our ideas into action. The computer, keyboard, visual display systems, vending devices, and research chamber all promised to be costly.

We submitted a grant proposal to the National Institute of Child Health and Human Development (NICHD), based primarily on our belief that our effort would yield a system that would benefit teaching and research with language-compromised children and young adults. We really were not very optimistic that we would be funded. However, one must try. We wrote the proposal and sported a team of specialists—a comparative psychologist (myself, Georgia State University), a psycholinguist (Ernst von Glasersfeld, University of Georgia), a computer programmer (Piere Pisani, University of Georgia), a developmental psychologist (Josephine V. Brown, Georgia State University), an electronics technician (Charles Bell, Yerkes Primate Center, Emory University), and a research technician (Timothy V. Gill). Throughout, Ms. Judy Sizemore (then Burns) would manage all office matters, a task that she did well then and, with me, for the next 35 years!

The grant application was submitted in the late fall of 1970; hence it went to NICHD through the Yerkes Primate Center, Emory University. My confidence that it would be funded was so low that I took another job in the spring of 1971, as Department Chairman of the Psychology Department of Georgia State University, within 15 miles of the Yerkes Primate Center. The move to Georgia State University was to become effective on September 1, 1971.

We were delighted and surprised to learn that NICHD would send a review team to Atlanta in the spring of 1971 to learn more about our proposal. Upon its completion I asked Dr. Keith Murray, chair of the review committee, what he thought of our chances. His countenance was not shining as he said, "Well, it isn't so hard to get the proposal approved. Most of them are. The problem is to get them funded." I saw no reason to cancel the planned move to GSU.

However, the proposal was approved and funded. We were really elated! They were excited about what we had proposed. Their excitement was contagious, and almost made me believe that apes might have untapped language skills that our system would cultivate and assay.

Our motivation was high, and we worked with a vengeance. The system's promise as a tool for teaching and research with children and young adults who, by reason of brain damage, had difficulties in acquiring and/or using language was more than enough motivation for our work, even if, as we feared, apes proved to have no language skills whatsoever.

We worked. We worked. We worked. We accepted a proposal to have Lana's chamber made of lucite plastic cubes, guaranteed to last 4 years—which, not coincidentally, was just how long our prized NICHD grant would support our work. (And yes, the chamber lasted 4 years, almost to the day.) The keyboard had been designed to permit relocation or scrambling of the keys so that Lana would have to learn about the lexigrams that embossed their surface, not just their locations. The lucite plastic chamber was installed in an embarrassingly small room in the nursery of the Yerkes Center. It was too small for comfort at 10×12 feet, but it was all that the center would allocate to us at the time.

The keyboard was meticulously installed. The computer was installed and programmed. We were set to go. It was January 1973. But what of the ape or apes to work in the system?

LANGUAGE AND SOCIAL INTERACTION

We decided to start with the chimpanzee, Lana, and an orangutan, Biji. The idea was that we would select the candidate who showed the most interest in the keyboard. To our chagrin, neither cared about it at all. Neither would respond to the keys so that they could be selectively reinforced by our computer-monitored keyboard system in order to teach language behavior. Well, Lana showed a bit more interest than did Biji. Translated, that meant that she bit the keys a bit more and looked at them somewhat more inquisitively. So we decided to bet on Lana. Biji was retired from the program.

Still, nothing happened. The dream of having an unaided ape acquire language skills via interactions with a computer-monitored system failed. Lana and we, the investigators, would all die before that happened. So what was to have been an automated language learning system was augmented with the introduction of a person, Timothy V. Gill, who was to work with Lana in a sensitive but demanding manner.

Through a combination of the keyboard, the presentation of the lexigrams to Lana, and Tim's tutoring, Lana began to learn. What did she learn? Initially she learned to press, in any order, a string of keys that we glossed as PLEASE MACHINE GIVE M&M PERIOD. Once she had learned to do thus, we required her to press each key, in turn, beginning with PLEASE and ending with the period key. Next the keys were scrambled in order, then placed in various locations on the keyboard to teach her to attend to the lexigrams on the surfaces of the keys, not just to their locations. Why the PLEASE? That was simply the key that signaled the computer that a request statement was being composed. Why the PERIOD key? That signaled the computer that the request had been completed and that now it was time to check the utterance for grammaticality.

What and how did grammar enter into the effort? Ernst von Glasersfeld constructed a system of grammar that was programmed by Piere Pisani into our humble PDP8 computer (the 8 stood for the kilobyte rating of its memory!). The details of the grammar are described in relevant chapters of a book[3] that recounts all aspects of the effort. Though the grammar was a finite state grammar—that is, what one could express with it was sorely limited—it nonetheless allowed us to assess whether Lana was sensitive to and master of the rules of grammar and whether she could learn the meanings of each of dozens of lexigrams. The grammar required, for example, that Lana write PLEASE MACHINE GIVE M&M PERIOD, PLEASE MACHINE OPEN WINDOW PERIOD, PLEASE MACHINE GIVE PIECE OF BREAD (or apple or chow, etc.) PERIOD, PLEASE MACHINE POUR MILK (or coke or juice, etc.) PERIOD, PLEASE TIM (or other person's name as relevant) MOVE INTO (or behind) ROOM PERIOD, PLEASE MACHINE MAKE MOVIE (or slide or music) PERIOD, and so on. Only those specific sentences would result in the consequences that we assumed were the incentives for Lana to work at the keyboard.

Slowly but systematically, Lana learned about her lexigrams and the constraints of grammar. Lana was very cooperative and benefitted rapidly from her training. Within 2 weeks she learned stock sentences which worked reliably for specific incentives and/or activities, specifically PLEASE MACHINE POUR JUICE PERIOD and PLEASE MACHINE GIVE PIECE OF CHOW (chow was a commercially made food high in nutrients). Furthermore, Lana soon showed us that she was reading what she was composing. This had to be the case because if she saw that she had made an error in the selection and sequencing of the lexigrams, she promptly hit the PERIOD key. This erased her mistake and cleared the keyboard so that she could start anew, which she quickly would do. The period key served as an electronic eraser for Lana! So Lana was sensitive to what she was doing. She monitored her work and controlled its quality.

READING AND SENTENCE COMPLETION

Lana was so impressive in correcting her errors in the formulation of sentences that we conducted a study in which we gave her sentence stems, some of which were correct and could be finished into one of several grammatically correct sentences, and some of which were flawed and could not be used to write a sentence. Valid sentence stems that Lana could build a sentence upon were, for example, PLEASE MACHINE GIVE, PLEASE MACHINE POUR, PLEASE MACHINE MAKE, PLEASE TIM, PLEASE TIM MOVE, PLEASE MACHINE, or just PLEASE. Examples of invalid stems are PLEASE GIVE MACHINE, PLEASE TIM MACHINE, PLEASE WINDOW, MAKE PLEASE, and so on. Very impressively, Lana was about 95 percent accurate in completing the valid sentence stems that we gave her, while rejecting stems that had fatal grammatical flaws.

We titled the article reporting this study, "Reading and Sentence Completion by a Chimpanzee." It was published by *Science*, and we were elated. At least for a time we were elated. Our joy was dampened by two letters to the editor. One letter claimed, in essence, "Well, now, we expect a lot more than this for us to say that a child is reading and writing!" Of course, children's reading and writing goes beyond what was accomplished by Lana, but we quite properly claimed that Lana had given us some evidence of reading and writing as operationally defined in the context of the study. There was nothing wrong with that conclusion. Nevertheless the two letters portended the future. We would have to contend with a number of critics, very few of whom were able to recognize sterling accomplishments by an ape.

CONVERSATIONS WITH LANA

But a few more notes on Lana's early progress should not be left out. We had hoped that someday we might be able to carry on conversations with Lana—not about the market or politics, but about various aspirations of the moment (getting a particular food or movie). We were understandably elated when it was Lana who initiated the first conversation! On March 6, 1974, Lana saw Tim drinking a Coke outside her lucite plastic room. In response to this view of Tim's activity, Lana composed a new sentence—LANA DRINK THIS OUT-OF ROOM PERIOD. She was allowed to do so. Once the Coke was gone, Tim and I talked and I reminded him that Lana knew the name of Coke. He should return and position himself once again outside Lana's cubicle. If Lana once again asked to drink "this" out of her room, Tim was to ask her, DRINK WHAT OUT-OF ROOM. Fortunately, Lana wanted more coke and asked once again if she could DRINK THIS OUT-OF ROOM PERIOD. Tim then replied, as per my request, DRINK WHAT OUT-OF ROOM? Lana responded, LANA DRINK COKE OUT-OF ROOM. Fantastic!! Incredible! Lana and we were on our way to an interesting future.

Another poignant conversation took place on May 6, 1974, when Lana asked for the name of something she wanted. Lana wanted a box, but at that time she did not know the lexigram for box. She wanted it because it held M&M candies. She asked for it by requesting, ?TIM GIVE LANA THIS CAN. Tim replied through his keyboard YES and gave her an empty can. That, of course, did not satisfy her. She then asked, ?TIM GIVE LANA THIS CAN. Tim replied that he had no can—it had been given to her. Lana then asked,

TIM GIVE LANA THIS BOWL. Tim then said "Yes," and gave her a bowl, an empty one. Then, as though asking for another person to help her, she typed, ?SHELLY. Tim replied, NO SHELLY. She was not there. Lana again said, as she had before, ?TIM GIVE LANA THIS BOWL. But before Tim could respond, Lana erased the sentence and then said, TIM GIVE LANA NAME-OF THIS. Tim replied, BOX NAME OF THIS. As though grateful, Lana replied, YES. She then continued, ?TIM GIVE LANA THIS BOX. And Tim gave it to her. Lana continued by requesting the name of a new vessel, a cup, and succeeded as before in using the new name.

Needless to say, we were impressed. Was her request merely chance? Perhaps. That said, she learned rapidly from Tim's response and used that information to solve her problem—to get the box and the prized M&Ms.

Lana seemingly loved Coke and became rather adroit in asking for it. ?YOU GIVE COKE IN CUP TO LANA; ?YOU GIVE COKE TO LANA IN CUP; YOU GIVE CUP OF COKE TO LANA, and so on across the years. In a similar way she asked for coffee and other prized drinks.

COLOR PERCEPTION AND ANSWERING QUESTIONS

Other creative work by Susan Essock entailed showing Lana hundreds and hundreds of Munsell color chips, noted for their precise colorations. Lana was asked to name each of them in turn as red, green, blue, yellow, and so on. Interestingly, she classified chips much as humans do. Lana named chips that portrayed hues with seemingly infinite gradation between colors. Her visual world was and is much like ours.

Lana's tutoring included work with a set of six objects each presented in six colors. So there were red, green, purple, yellow, black, and white shoes, bowls, cups, balls, boxes, and cans. Lana learned to give the color or the name in response to specific requests. When presented with sets of objects (>1), she could be asked for the color of a specified object, for instance CAN or for the name of an object specified by color, for instance PURPLE. Her accuracy level was better than 90 percent even if the sets presented consisted of up to four items, for example, a red can, a purple shoe, a black cup, and a green box. The time required for her to answer the question increased as the number of items in the set increased; that was expected because she had more items to search among before she could provide its color or name. We were very impressed by Lana's performance in this, as well as in all other, tasks.

DON'T SAY IT ISN'T SO

The first report that challenged any suggestion that apes had productive syntax was published by Terrace, Petitto, Sanders, and Bever in 1979.[4] Their study with their chimpanzee, Nim, had an extraordinary impact. For Terrace, the two basic requisites for language are semantics and syntax. The semantic bases of the signs were inferred by assessing the appropriateness of Nim's signs in response to objects and novel events; however, no controlled tests of these skills have been reported.

A detailed analysis of Nim's 20,000 recorded utterances was made to assess whether there was evidence of syntax. The conclusion was that there was none. Although there were preferential sequences of signs, there was no evidence that the information content of multiple-sign utterances was greater than that of single-sign utterances. Many of Nim's utterances were either partial or total imitations of his teacher's statements, enriched by a small number of preferred signs that increased emphasis, but did not increase the substance of the message. Terrace reviewed available materials from other signing projects with apes and found evidence to support his contention that the same factors operated in them. The conclusion was pointed: the signing apes could not construct a sentence. They were only simulating language because their behaviors only simulated syntax. Nonetheless, the fact that Nim gave no evidence of syntax does not mean that apes are incapable of syntax. Bear in mind two things. First, Nim was still a very, very young chimpanzee when his language schooling at Columbia was terminated. At that time, he was not quite 4 years old. Competence for discourse-governed syntax, *if* extant in the ape, may not appear until adulthood, and probably requires *some* specific training, for which Nim had none. With only one-third of the cranial capacity of *Homo sapiens*, the ape must have everything going for it in terms of maturation and training if it is to give clear evidence of syntax. Second, no evidence was reported that Nim's signs had referential meaning for him. Although at times Nim signed in a manner appropriate to the context, there were no tests to assess whether these signs had referential or semantic meaning. Without referential meaning, Nim could not be expected to use signs with syntactic competence in discourse.

When Lana was $5\frac{1}{2}$ years old, her computer system was expanded to allow her to construct sentences of up to 10 lexigrams, as opposed to the earlier limit of 7 lexigrams. Lana produced 36 novel sentences of at least 8 lexigrams during the first 24 days after the computer system was expanded. Lana's functional lexigram vocabulary at the time was >100 symbols. An analysis of the lexigrams in these 36 sentences showed that Lana was the first to use 92 percent of them in a given conversation. Only 8 percent were used first by the experimenter. Thus most of what was produced in long sentences was original, at least in the 20-minute time frame that was examined. Lana was seldom if ever imitating the utterances of her caretakers (some words would naturally recur in a conversation, and 8 percent does not seem an unexpected percentage in the absence of imitation). Her sentences were unique to that time frame, and they were initiated without prompting by the humans communicating with Lana. In addition Lana, unlike Nim, seldom repeated words within an utterance. Though Nim often imitated his teachers, Lana did not.

Terrace's work was, unfortunately, often taken as proof not only that apes have not demonstrated any evidence of true language, but also that they are incapable of language! Never in the history of behavioral science has a single report to the effect that a phenomenon (that is, syntax, in this case) had not been observed (that is, negative results) been accepted so broadly as a negative proof—in this case that the ape has no capacity for language and that there was no longer enough justification for studies of apes' potential for language.

Research into questions about apes and their language skills nearly died in the wake of Terrace's report. Despite frequent and clear denials by the majority of researchers in the ape language field, the question that was often assumed to justify ape language research was "Do apes have the capacity for human language?" The real question was what language skills apes could master. Thus, in the view of most ape language researchers, Terrace's position was most unfortunate and indeed unfair, for it raised beyond reasonable bounds the criterion for justifying ape language research.[5] Given this extraordinary requirement, coupled with lack of a generally accepted working definition of the criteria to be met for the conclusion, "Apes have language," confusion and despair were just a matter of time. Desmond[5] commented,

> Then there was also the problem of uncritical acceptance of reports by ape-buffs—those who would have the apes given their rights and status as the free men, which they are not.... The overriding urge to assist apes in giving man his comeuppance, as if this were a Darwinian imperative, has boomeranged to the detriment of the ape, who is now judged according to an impossible human standard which should never have been set.... The reason

that Terrace's conclusion comes as a profound blow is because of this monstrously misplaced standard which has set the tone of the ten-year debate…Who suffers? But the ape? The cry will now be "They don't have human language after all." But it might as well be "So they aren't human after all," for all the sense it makes. (pp. 49–50)

CARRY ON!

Though a great deal had been learned about language and about apes, the future of ape language studies looked bleak in the immediate post-Terrace era. But the field did survive and began to thrive, buttressed by the results of a notable study conducted by E. Sue Savage-Rumbaugh. The study was designed to answer seemingly impossible questions like, "Are the lexigrams meaningful to the apes? What do the apes think about, if anything, when they see a lexigram? Do the lexigrams function as symbols that represent things that are not necessarily physically present?" The study was as follows.

Representational Symbols

Work with two chimpanzees, Austin and Sherman, had repeatedly shown that it was possible for chimpanzees to learn symbol-object associative responses without learning symbol-object representational responses. Accordingly, Dr. Savage-Rumbaugh designed a test that would reveal whether or not a particular lexigram was functioning at a representational level. The test demonstrates first *that the chimpanzee can use symbols to represent items not necessarily present, and second, that the chimpanzee conceptualizes and categorizes things in a way that includes use of the learned symbols (word-lexigrams).*

This testing was performed in three stages. Each stage had a training phase and a test phase. The same items were used during all the three training phases, and novel items were used during the test phases.

Sorting and Labeling Real Objects

Sherman and Austin were first taught to sort six exemplars into two bins—three foods and three tools, for which they had learned specific lexigram names. Next, again with the same three foods and three tools, they were taught to use the generic lexigram symbols FOOD and TOOL at the keyboard as each exemplar was presented in turn. A transfer test with other foods and tools once again revealed high-level, accurate naming as either food or tool.

Labeling Photographs

Once the chimpanzees could accurately classify virtually any item as a food or tool (all nonedibles were classified as tools), the task was made more abstract by replacing the real items with lexan-encased photographs. Now none of the exemplars were to be eaten; furthermore, they all tasted alike. The only difference between the tool and food photographs was their appearance. The classification decisions had to be based on recalled, not present, characteristics.

Labeling Lexigrams

During the third phase, the training photographs were replaced with the lexigrams that stood for the various training objects. Again the chimpanzees were able to classify, only now they had only a symbol to view when making their decision.

Training in this phase began as it had in the previous ones. The original group of training foods and tools was used once again. This final phase of the study was the most critical of all, for it alone was unequivocally a test of the referential value of the symbols. This test included the following critical constraints: (a) *it required a completely novel response, one never before given by the chimpanzee or by the experimenter when working with the chimpanzee;* (b) *it was administered under blind conditions with the experimenter out of the room; and* (c) *it required that the chimpanzee use one symbol to classify other symbols.*

Remarkably, Sherman categorized the novel lexigrams correctly on 15 of 16 trial-1 presentations, and Austin categorized them correctly on 17 of 17 trial-1 presentations. It was concluded that the chimpanzees had to refer cognitively to the specific referent of each symbol and, based on the recalled characteristics of that referent, assign it to a class of functionally related items through use of another symbol.

It seemed clear then that these chimpanzees were able to use their lexigrams as private representations or symbols, and that categorization is one and the same for the physical objects and for the lexigrams that stand for them. This demonstration is very important for a variety of reasons: (a) it attests to the chimpanzees' ability to manage symbols and concepts, and (b) it supports the conclusion that our language data are not simple associations but are, rather, reflections of cognitions and understandings of relationships.

Syntax

Regarding syntax, Lana not only became proficient in ordering lexigrams in accordance with Yerkish grammar, but also demonstrated some ability to insert lexigrams and use paraphrases.[6,7] Even Lana's incorrect sentences were in the majority not random productions. Our work with Sherman and Austin demonstrates that chimpanzees *can* use learned symbols referentially. They also used the symbols singularly to achieve communication between themselves and with humans regarding tools, foods, and locations that otherwise exceed their unlearned nonverbal communication skills. The work with Lana clearly demonstrated her ability to learn to chain word-lexigrams into strings that simulate sentences generated by humans. Many of her productions were competent and conveyed novel meaning. What is needed, then, is the integration of these skills—Lana's on the one hand and Sherman's and Austin's on the other. The gap has yet to be spanned.

WHAT'S IN A WORD?

One way of deciding whether or not a given unit of behavior (key pressing, motoric hand displacement, vocal articulation, etc.) is functioning as a word is to determine whether that behavior can be used reliably to transfer information from one party to another. The unit of behavior must also function referentially, regardless of whether or not its referent is present.

For the chimpanzee, if not for the human child as well, word acquisition per se is not an all-or-none phenomenon, in the sense that a word is either known or not. Rather, there are various levels of "wordness," or competence with particular symbols; these reflect both the chimpanzee's comprehension of a given symbol and its communicative symbol-use in general. Simple usage or production criteria prove to be inadequate to distinguish various levels of comprehension. Such criteria are also inadequate for determining the chimpanzee's comprehension of the symbols when produced by others.

LONG-TERM MEMORY FOR LEXIGRAMS

More than 20 years after she last worked with lexigrams for various foods, colors, and objects, Lana identified a majority of items with the correct lexigram on the very first test trial.[8] In the food category, she remembered the lexigram for apple; for the colors, she remembered black, blue, orange, and yellow (four out of six colors); and in the object category she remembered ball, box, can, cup, and shoe—and forgot bowl. She still knows and uses a relatively large number of lexigrams. Interestingly, Lana has continued to learn new lexigrams, despite no longer being trained to use them.[9] Her learning of new lexigrams is attributed to her long-term observations of other apes' and humans' use of them and the lexigram's usefulness in getting to go outdoors, see things of interest, and so on. Observational learning in this situation has allowed Lana to gain increased control over events in her life through the use of symbolic communication.

LANA COUNTS

Lana's skill with numbers has as long a history as her importance in language acquisition research. In her earliest years, Lana showed us that chimpanzees are very sensitive to even slight differences in quantity, as she was able to select the larger of two quantities of cereal pieces when given a choice. Ratios containing quantities of one to five items including equal ratios (to determine whether there were side preferences) were presented to Lana. Lana's performance, in general, was very good, and significantly better than chance. When ratios involving larger numbers were included (up to 9 : 10), Lana's performance deteriorated somewhat, particularly when the difference between the two quantities was only 1 or 2 items. However, her discriminations were still very good. In another experiment Lana had to master MORE and LESS labeling. Washers of various sizes were presented to Lana. She was then asked WHAT MORE or WHAT LESS by caretakers using the computerized lexigram apparatus. The quantities one to five were used in all possible ratios. Lana had to produce her answers in the following way: point to the question, select the appropriate lexigram answer, correctly label the amount referred to in the question as MORE or LESS, and tap each washer placed along the line on the labeled board. Lana was 89 percent correct over the 100 test trials. She had some difficulty distinguishing the amounts that differed by only one item (with 3 : 4 and 4 : 5 being the most difficult discriminations), but her performance again indicated chimpanzee sensitivity to numerousness.

Lana's work with numbers did not end there. In Phase 1 of a computerized paradigm, Lana had to learn the relation between a target Arabic numeral and the subsequent production of a number of responses related to this numeral. Lana had to contact a numeral and then select boxes from the bottom of the screen in the correct order until she indicated that she was finished. She did best in counting to 1, then to 2, and 3. Toughest of all was to count to 4—but she did so with greater than chance accuracy.

CROSS-MODAL PERCEPTION

Cross-modal studies involve recognition of objects first sensed in one mode (for example, visual) in another mode (for example, tactual). Prior to 1970 it was generally believed that cross-modal perception was uniquely human and dependent on language. In studies of cross-modal perception, a common method has been to use the matching-to-sample procedure. It permits immediate or delayed comparison of stimuli in two modalities. Generally, extensive training with nonhuman primates has been necessary for them to learn to select one of several stimuli that is identical to the sample. For example, if a cup is the sample, the correct choice would be another cup, presented along with, say, a box and a spoon. Eventually, primate subjects can become so competent that they do well on a series of problems, each presented for a single trial. Next, cross-modal or inter-modal perceptual competence can be tested. For instance, an ape might see one object, then select by touching another of its kind from a selection of objects presented in a container into which the ape can reach but not see. Cross-modal perceptions of equivalence or difference between objects requires representations that are modality-independent. If an ape can learn names of objects, it should facilitate its cross-modal perceptual accuracy, as it is known to do in children's cross-modal perceptions.

In cross-modal studies Lana was first taught to declare whether pairs of items presented to her were comprised of identical or different items. We intended that for the identical pairs, she would use the lexigram for SAME, and the lexigram for DIFFERENT when they were not. She used the SAME lexigram with no problem, but refused to use the DIFFERENT one. Rather, she coined her own answer for pairs of different items—NO SAME. It took Lana only 29 trials to be able to name accurately each of 6 objects for which she had names on the basis of touching but not seeing them. In the cross-modal task that followed, Lana could see only one member of each pair. The other item she was permitted only to feel in a bag. Could she determine sameness or difference through vision and touch? Yes! On 60 trials, she was 92 percent correct. The items used were from the set of bowl, box, shoe, cup, ball, and can referenced above.

These items were, of course, familiar and named. How would she do with familiar objects that had no names? Lana was given 60 items to view and to manipulate, but she had no names for them. After two weeks of such experience, she was tested with them and with other 60 items as well—ones that were not even familiar! With the familiar but nameless items Lana was 80 percent correct, but she was only 57 percent correct with unfamiliar objects. Familiarity facilitated cross-modal perceptions of sameness and difference. Even more effective, however, were names! This finding was corroborated in subsequent tests with pieces of food, where one piece could be seen and the other piece not seen but palpated. If the pieces of food were familiar but without names, so far as Lana was concerned, she did less accurately than if the foods were named. Clearly, symbols helped Lana in her cross-modal perceptions. This evidence strongly indicated to us that for Lana, lexigrams served as symbols for objects.

AN OVERVIEW OF LANA'S LESSONS

An important finding in our studies with Lana was that she exhibited many significant linguistic skills for which she had received no specific training. Unquestionably she would never have acquired them spontaneously had she not received specific training in certain language fundamentals; nevertheless, having received such training, she expanded her ability to use her symbols and newly acquired communication skills in a number of significant and novel directions. All of her innovations appear to reflect a substantial ability for abstracting and generalizing. The data show that Lana's performances on occasion involved transferring information learned in one situation to another, and generalizing skills learned in limited contexts to deal with broader problems. Her behaviors are fine instances of emergents (Rumbaugh, Washburn, and Hillix, 1998).

The first instance of Lana's untutored acquisition of linguistic-type skills—her learning to read lexigrams—was the most crucial. We had taught her to read the lexigrams embossed on the surfaces of keys, but we had not implemented methods to train her to equate these lexigrams with the lexigram-facsimiles produced by and on the projectors. One of our greatest concerns was that she might never attend to what was shown on her projectors or grasp the significance of the order in which the lexigrams were projected. Without these skills, the type of communication we planned simply would have been impossible. Our fears were groundless, however. Without specific training, Lana promptly read the projected lexigrams, and she also grasped the importance of their order to the point that she could complete valid sentences begun for her or erase invalid sentence beginnings presented to her by the experimenter. She did all of this without specific training.

Lana also used names in novel ways with no specific training or encouragement. Though her initial and specific naming training was arduous, she soon seemed to grasp the basic concept that things have names. She became quite facile in learning the names of things, and, as in the incident with the box, sometimes asked what the name of something was—then asked for it by name. She combined lexigrams smartly to ask for certain things that had not been named for her (for example, a cucumber was named by Lana the BANANA WHICH-IS GREEN and an orange fruit requested as the BALL WHICH-IS ORANGE (that is, colored) or the APPLE WHICH-IS ORANGE, and so on). In addition, she spontaneously began to use THIS as a pronoun to refer to objects whose names she had not yet learned.

Lana's use of NO is a particularly interesting example of her generalizing ability. Having learned this word in the context of simple negation, she expanded it with no particular training to use as a form of protest.

Lana's extended use of stock sentences was quite creative. For example, the specific sentence, PLEASE (person's name) MOVE BEHIND ROOM PERIOD was taught so that it might be used by Lana to get a caregiver to carry food back behind her cubicle so that it could be placed in the appropriate vending device for access. Lana extended its use to get caregivers to move behind the room so that she might bring their attention to the fact that one of the vending devices was jammed. This seemed clear because, after having succeeded in getting the person to MOVE BEHIND ROOM Lana then would ask the machine to vend something specific, such as PLEASE MACHINE GIVE PIECE OF APPLE PERIOD then point to the vending device to ensure that its malfunctioning had been seen!

How was Lana able to do thus? It is my own suspicion that apes' intelligence includes a covert ability to generate idiosyncratic symbols that in turn enables the ape to do highly intelligent problem solving and cognitive operations. Yet the apes are not clever or intelligent enough to even try to reach social agreement upon symbols and public referencing or use of them. Hence, they do not develop a language system on their own.

The merit of our lexigrams and computer-monitored keyboard perhaps was to enable Lana to learn through experience and social discourse with us about symbols' meanings. Thus it became possible for us to share an *a priori* vocabulary with Lana, and grammar as well. Lana could then step across a threshold into entry-level language operations. She loved her work and gave happy vocalizations whenever her keyboard was turned back on after many hours of being down or off for maintenance. She knew the value of the keyboard. It enabled her to enter the world of language otherwise denied to her.

Lana's competence with language was perhaps illustrated most clearly when she got into what seemed to be both a playful and a testy mood. In the fall of 1976, for example, a Russian delegation visited the Yerkes Regional Primate Research Center. They were very interested in seeing Lana. The Cold War tensions were still present at the time of their visit.

Tim wanted Lana to look good in her skills, so he started the demonstration with a very simple stock question with which Lana was thoroughly familiar. But Lana failed to cooperate, instead asking him to do things for her! Tim: ?LANA MAKE WINDOW OPEN. (The question mark was always used at the beginning of a request statement in the Yerkish grammar.) Lana responded: LANA WANT TIM GIVE M&M. Tim replied: NO. Lana then said: LANA WANT TIM GIVE M&M TO LANA. (This is actually a paraphrase of her first statement in this exchange.) Tim replied: NO. TIM WANT LANA MAKE WINDOW OPEN. Lana replied: NO. Tim insisted: YES. Lana started to respond, YOU,. —.but was cut off by Tim, who next insisted, LANA MAKE WINDOW OPEN. Lana replied, YES, but did nothing! Again, Lana started ?YOU—, but Tim cut her off and commanded her, MAKE WINDOW OPEN. Lana this time replied, NO. Tim insisted, YES. Lana next requested, LANA WANT TIM GIVE CUP OF JUICE. NO, responded Tim, TIM WANT LANA MAKE WINDOW OPEN. Lana defiantly responded, NO. LANA WANT TIM MAKE WINDOW OPEN. On this occasion, Tim gave in and typed in the sentence, PLEASE MACHINE MAKE WINDOW OPEN.

But Lana got nothing—neither her M&Ms nor juice. The Russians laughed with great pleasure. Lana had won the contest. The Russians then quipped that everything was all right and there was no need for anyone to construe the exchange between Tim and Lana as the occasion for yet another international incident!

FAREWELL TO LANA!

It is not too late for us to be reminded of Lana's genius and her bold ventures into the brave new world we call language. The system and methods used successfully with Lana afforded limited language to children and young adults who could not master any language by more conventional methods. The achievements of our computer-monitored keyboard system validate the conclusion that Lana learned about language. She read the lexigrams produced on the row of projectors above her keyboard, initiated goal-directed conversations, became sensitive to her errors at the keyboard and erased them, built upon sentence stems that we gave her as though she had started them, learned what to do in various cross-modal tasks, and executed them with competence. Her mastery of lexigrams appeared to help her in those cross-modal tasks; all of this supports the conclusion that Lana had learned language skills that helped her adapt to the challenges of her world in ways that she never could have without those skills.

ANIMAL LANGUAGE RESEARCH MEETS THE COMPUTER

What follows is my (William A. Hillix's) commentary on some of Rumbaugh's work, derived from his published reports. In Chapter 6 of his book-length report of the language analog (LANA) project,[3] Pierre Paolo Pisani reports on the development and nature of the computer programs that were developed so that Rumbaugh could live out his dream of computerization. Chapter 7 by Harold Warner and Charles L. Bell describes the physical system that interfaced the chimpanzee and the computer.

Gordon Hewes contributed a chapter, "Language Origin Theories," a conceptual and historical overview of opinions about language origins. He identifies an amazing number of general theories on the subject, his favorite being the gestural theory in combination with some other (because a gestural theory cannot account completely for the development of vocal language). An appealing alternative is the mouth-gesture theory, according to which vocalizations tend to accompany actions of hands, including hand gestures. The ancients had surprisingly sophisticated notions about language, and several religions like to claim that the original language was theirs—Hebrew if the claimant is Jewish, Persian if Persian, etc. An ancient Egyptian king was alleged to have had two children raised without human language contact in order to settle the problem of native language, and perhaps to study whether language developed inevitably or because our human nature required an appropriate nurture. Hewes also discusses the possible limitations of animal language; although he reaches no conclusion, he favors the view that there is a continuum of language potential from humans down through the other great apes and primates.

DEVELOPING AN ARTIFICIAL LANGUAGE

Although the development of technology and a broad view of language were critical to Rumbaugh's endeavor, the central element in the study had to be the design of a language for Lana and LANA. Ernst von Glasersfeld describes what he thinks language is and how the lexigrams were designed in his chapter, "Linguistic Communication: Theory and Definition." He alludes to the debate about the definition of language and suggests criteria for defining language (versus speech, which the reader knows well by now is not the same). Von Glasersfeld correctly points out that semantics is the *sine qua non* of language—ergo, grounding "symbols" in experience is absolutely the key element. He alludes to the "mathematical theory of communication" (Shannon) and to information theory (MacKay and Wiener). Von Glasersfeld, as a linguist, is naturally aware of the "design features" that Hockett used to describe human language. Von Glasersfeld claims that "purpose" is essential to language, and has to be present in the sender. He adopts the cybernetic model of communication; in this model, a thermostat "has a purpose" because it "aims" to achieve a preset temperature by using feedback from the present temperature to "decide" whether or not to keep the heater (or air conditioner) operating. A thermostat objectifies what it means to have a purpose, an intention to achieve a goal. However, a thermostat presumably has no awareness of its goal, or of its own operations, things that a human with a purpose presumably has. The thermostat also lacks any reference to the future, a reference that is the usual property of human purposive acts. Von Glasersfeld said that "in order to be communicatory rather than inferential, signs have to be in some sense artificial." In accordance with that belief, the lexigrams designed for Lana to use were strikingly artificial. It is much easier to learn more iconic symbols, and even a photograph of an object has a degree of artificiality. However, the completely artificial lexigrams actually designed for Lana presented her with symbols that, if mastered, would provide a convincing demonstration of her ability to learn a true language.

Von Glasersfeld says that language requires three things: first, a lexicon (which can be small), which is used symbolically. Von Glasersfeld says this is similar to Hockett's design feature of displacement; that is, to quote Hockett,[3] "We can talk about things that are remote in time, space, or both from the site of the communicative transaction" (p. 64). Second, symbols have to be *detachable* from immediate input; the symbol is a symbol because it can be used regardless of contextual situation, in propositions, suggestions, or the like. Bee dances, says von Glasersfeld, are communicative, but they lack this property because they do not occur except after return from a food source. (The bee dance is, however, artificial because they are not flying, and the direction is not literally the same anyway.) The final, third property of a language is that its elements can be combined to create new meanings (that is, the language has grammar). This is equivalent to Hockett's openness or productivity.

Von Glasersfeld designed the language very logically; he wanted to make each design element correspond to some grammatical category. The colors of the lexigrams corresponded to rough semantic classes; for example, any violet lexigram represented an autonomous actor, and any red lexigram represented something edible. Nine different shapes were presented in various combinations to make discriminable lexigrams.

Von Glasersfeld describes[3] the classes of lexigrams; he says his language allows for 46 lexigram classes, "...of which 37 are in use at the moment" (p. 97). Example of the classes of lexigrams included familiar primates, unfamiliar primates, nonprimates, inanimate actors, absolute fixtures (cage, piano, room), relative fixtures (window, door), transferables (balls, etc.), parts of the body, edible units, locational prepositions (in, on, etc.), additive conjunctions, ingestion of solids, ingestion of liquids, relational motor acts (groom, tickle), locomotion, etc., etc.

Von Glasersfeld describes how the parser works by indicating the relationships into which a given lexical item can enter. Yerkish works with some 30 correlators that connect types of items; for example, the first one connects autonomous actors performing a stationary activity, as "Lana drink." Another example is a movable actor changing places, as "Tim move." Yet another example is predication, as "Banana black."

Yerkish has only active voice, and 3 moods: interrogative, indicative, and imperative. The parser responded to strings that were correct in Yerkish, and did not respond to others (at least generally).

LANA USES LEXIGRAMS

Results as of 1974 indicated, according to von Glasersfeld,[3] that Lana had a "strong tendency toward grammaticality" (p. 128). She was, of course, rein-forced for grammatical strings, and not reinforced for non-grammatical strings. Von Glasersfeld believed that the grammaticality of Lana's "utterances" was caused primarily by the design of the language, which was intended to reflect the conceptual structures common to humans and chimpanzees. Thus Lana expressed her representations of situations through the choice and ordering of her lexigrams. He thought that Lana learned "rules" of grammar that "are relatively close to the rules that govern conceptual representation" (p. 129).

After Lana had used lexigrams extensively to make requests, she was trained to name objects. It took 1600 trials for her to learn to name M&Ms versus banana slices, but she eventually learned to name them, and later transferred the naming notion to other incentives and objects in a relatively small number of trials. She apparently got the idea of naming at some point, and named slides with no specific training.

At $4\frac{1}{2}$, Lana was tested for her conversational abilities by telling her lies or presenting problems to her. She was asked for morning decisions in the after-noon (What want drink?) and vice versa, or shown desirable but unavailable foods or drinks. She nearly always engaged in conversation until she got some-thing acceptable, and the conversations were, as expected, longer when her expectations were contradicted.

One of the practical benefits of this research was a conversation board that was used with handicapped children. A number of children with no productive language learned several words when they were given a lexigram board and taught to use it. As I have said earlier and will say again, all of the techniques used with apes—sign language, plastic symbols like those used by Premack, and lexigrams have helped handicapped human children or aphasic adults.

INTERPRETING ANIMAL LANGUAGE BEHAVIORS

Duane Rumbaugh and Sue Savage-Rumbaugh[3] presented an overview of the relationship between action, communication, and language as a conclusion to the report on the LANA project. They argue that responding to a stimulus (for example, the fruit looks ripe, the dominant male looks mad) is not really communication. However, all types of information uptake are on a continuum, certainly with a continuum between intentional displays of dominance and symbolic indications of the same thing. Displacement in time and space of the symbol from the referent is said to be a, or the, main dimension of language, although of course the writers do not deny the importance of rules and syntax.

Another important point is the continuity between different types of cognitive competence; when competence is sufficient, language becomes possible. Some cognitive operations become possible through interaction with the environment. Human children attain this competence, perhaps sometimes in ways analogous to the ways in which chimpanzees acquire such competence. Researchers should make a great effort to identify as many cognitive operations underlying language as possible; Chomsky's transformational grammar is such an attempt, but the operations he suggests seem too complex to underlie primitive languages. Simpler abilities, some already much studied by psychologists, are clearly necessary but perhaps not sufficient for language to emerge. Among them are the abilities to discriminate, associate, produce, and evaluate sequences, attend, observe the sender or recipient, and map situations into symbols and symbols into situations. These two mappings are not the same, and some of the results with Lana demonstrate this lack of symmetry. Before she received specific training on naming, Lana could request what she wanted, but she could not name the object that she asked for.

Rumbaugh and Savage-Rumbaugh also noted that the role of context is critical. They believed that the cold quantification of information theory is not going to help much—that we need to know what the organism wants to communicate and study its behavior. In addition to studying context, researchers need to study more carefully the relationships between the verbal and nonverbal elements of communication.

These authors propose a hierarchy of information-bearing systems, from least to most linguistic: physiological attributes (female has sex swelling), social interchange (rough versus gentle play), elaborated social patterns (dominance displays, an instance of true communication), iconic gestures (beckoning to "come here"), and, finally, arbitrary signs (true language). One can quibble about details, but such a classification system provides a useful structure.

Information about the latest research and the status of the Language Research Center and the Sonny Carter Laboratory is available at their web sites, www.gsu.edu/lrc and http://www.gsu.edu/~lrcdaw/b2ec2.htm.

REFERENCES

1. Language and Man (1975, September 30). *The New York Times.*
2. W. H. Furness, Observations on the mentality of chimpanzees and orangutans. *Proceedings of the American Philosophical Society,* **55**, 281–290 (1916).
3. D. M. Rumbaugh (Ed.), *Language learning by a chimpanzee: The Lana project* (Academic Press, NY, 1977).
4. H. S. Terrace, L. A. Petitto, R. J. Sanders, and T. G. Bever, Can an ape create a sentence? *Science,* **206**, 891–902 (1979).
5. A. J. Desmond, *The ape's reflexion* (The Dial Press/James Wade, NY, 1979).
6. H. F. W. Stahlke, On asking the question: Can apes learn language?, in: *Children's language,* edited by K. E. Nelson (Gardner Press, NY, 1980), Vol. 2.
7. J. L. Pate and D. M. Rumbaugh, The language-like behavior of Lana chimpanzee: Is it merely discrimination and paired-associate learning? *Animal Learning and Behavior,* **11**(1), 134–138 (1983).
8. M. J. Beran, L. Pate, W. K. Richardson, and D. M. Rumbaugh. A chimpanzee's (*Pan troglodytes*) long-term retention of lexigrams. *Animal Learning and Behavior,* **28**, 201–207 (2000).
9. M. J. Beran, E. S. Savage-Rumbaugh, K. E. Brakke, J. W. Kelley, and D. M. Rumbaugh. Symbol comprehension and learning: A "vocabulary" test of three chimpanzees (*Pan troglodytes*). *Evolution of communication,* **2**, 171–188 (1998).

CHAPTER TEN

A Cultural Approach to Language Learning

THE LANGUAGE RESEARCH CENTER (LRC)

The Language Research Center (LRC) where Drs. Duane Rumbaugh and Sue Savage-Rumbaugh work is a scatter of buildings on 50 acres of gorgeous Georgia pines within the boundaries of Decatur, Georgia. Decatur is a suburb of, and continuous with, the great southern city of Atlanta. Within these buildings are dispersed several chimpanzees, at least seven bonobos (formerly called pygmy chimpanzees) and several rhesus macaques. Up a rocky, sometimes muddy, road along the borders of this Georgia State University property is a wooden house that seems virtually to grow out of the trees that surround it. You always need a high clearance vehicle to get there, and when it rains you had better have a 4-wheel drive. Sue lives there, and Nyota, a 4-year-old bonobo, has been living there with Sue almost as often as with his bonobo mother, Panbanisha.

Most of Dr. Savage-Rumbaugh's work these days is with bonobos like Nyota. Bonobos, whose proper species designation is *Pan paniscus*, split off from the human line only a few million years ago, probably at the same time as chimpanzees, *Pan troglodytes*. For a time bonobos and chimpanzees shared a common ancestor. Then not so much later, about 1.5 million years ago, bonobos split from chimpanzees. Most scientists now insist on calling *Pan paniscus* bonobos rather than pygmy chimpanzees to emphasize their separation. Kano[1] has studied bonobos for years, and finds that, unlike chimpanzees, male bonobos do not kill each other, and infanticide is rare. They often have face-to-face sex, and females engage in genital rubbing with other females. Bonobos generally seem more sociable and less aggressive than chimpanzees.

151

They live south of the Zaire River in an area richer in food sources than the north of the Zaire, where chimpanzees live. They may have evolved more slowly or less competitively than chimpanzees because of their kinder environment.

Dr. Savage-Rumbaugh sometimes allowed visitors for Nyota at her house when he was very young, and I was one of those guests as part of a late-1999 trip to Atlanta. When I arrived, I was delighted to see her sitting on a couch with Nyota, who was then only 18 months old. Nyota's father is P-Suke, a fairly recent gift to the LRC from supporters in Japan. Panbanisha is Nyota's mother, but was willing to share her son with Sue; it is not unusual for bonobo mothers to allow their infants to visit other bonobos or acceptable humans.

This was my first opportunity to interact directly with a bonobo, and I was delighted to interact with Nyota. Nyota and Sue seemed to sense my pleasure. Sue told me to let Nyota approach me, rather than trying to approach him, and his curiosity soon began to get the better of his hesitation to approach a stranger. He came over for a few visits, but soon retreated to Sue's side. Then she said it was okay to come and sit beside them, and I did.

Nyota was even more at ease with his security blanket, Sue, at his side, and started to interact with me pretty freely. It is an exquisite experience to have Nyota come peer into your eyes from a distance of about 4 inches! Nyota is even more spontaneous than a human child, and is also much more precocious. At 18 months he got around with great facility on either two or four limbs, and walked up a tree almost as easily as he walked on the ground.

When Nyota looks curiously into your face you get the feeling, however misleading it may be, that you are staring back at your own past. At the same time, the openness and honesty of the young animal gives you a fresh look at the modern world.

The following afternoon I was able to continue my conversation with Sue at the LRC's main building, despite the fact that she was terribly busy planning for a talk that she was to give a few days later at Kyoto University in Japan. Among the many things she told me was that she has not, like most people, given up on teaching bonobos or chimpanzees to speak a vocal language that humans can understand. She thinks that the human teacher should be expected to adapt at least as much to what the chimpanzee is able to vocalize, as the chimpanzee is expected to adapt to the requirements of human communication. Because so few humans understand sign language or other "artificial" methods of communication, most of us cannot evaluate, on our own terms, how well animals are able to communicate. In this case hearing would be believing, even if the sounds required considerable training to understand.

After the interview, I did not expect to see her again. However, I was pleasantly surprised the following day to receive a warm invitation to come visit her and Nyota. I was sure that the invitation came because Nyota and I had liked each other so much; I was hired for another visit, in much the same way that Washoe hired Roger Fouts after he thought he was having a bad interview with Allen Gardner. So Duane Rumbaugh, who was taking care of me, took me back to the LRC. Sue and Nyota showed up soon after we arrived, with Nyota in an infant stroller.

Nyota and I played chase for about 45 minutes. At first I exclusively chased Nyota; Sue pointed out that I was letting him dominate the game, and suggested that I try to turn it around. So I hid behind a tree and waited for him to chase me, which he quickly did. Nyota's gross motor skills appear to be as good as those of an adult human. He readily climbs trees, ambulates rapidly on four limbs or two, can jump into the air while standing on two feet, and so on. He hung from the limb of a small tree by one hand and kicked me on the head (he is very strong, and can kick hard for a 25-pounder). As to the stroller, he jumped in when he wanted a ride and jumped out and ran when he thought Sue was not wheeling him fast enough.

After about 45 minutes the evening was getting cool, so we went inside, and Sue put Nyota on a long conference table. I sat on one side, and Sue on the other. Nyota was extremely active, moving back and forth between Sue and me, and trying to go to the end of the table and approach my son Gaines, who had accompanied us. Gaines had not had a TB test recently, so Sue had to stop Nyota with repeated and vehement "No's," which he did his best to ignore. She sometimes had to threaten him physically. He used every stratagem he could devise to get his way, but nothing worked. Nyota was generally reasonably obedient, despite his high activity levels, but on one occasion he was forbidden to get onto the floor and did it anyway. A brief chase by Sue and me ensued, and we caught him and put him back on the table.

Nyota and I played several games on the table, with him moving up and down the table either bipedally or quadrupedally. When he lay down on his stomach facing Gaines, I would grab a leg and slide him down the smooth table on his hairy stomach, which he seemed to enjoy; he gave me repeated opportunities to give him that free ride. Periodically he came over to give me a hug and a kiss. That is an experience with no parallel in my life, and if you like animals as much as I do, it would give you, as it did me, an ultimate thrill.

Early in the session, Sue gave Nyota some grape jelly out of a jar and left the lid on the table. Some time later I moved the lid around while looking at Nyota, and he immediately picked up on the opportunity for a game of keep-away. I would move, he would grab, and sometimes I would let him take hold for a quite equal game of tug-of-war. His strength is astonishing, and I was reminded of what Winthrop Kellogg said about Gua, the female chimpanzee that he and his wife kept with their son, Donald, for 9 months. Dr. Kellogg said that Gua's muscles when relaxed were as hard as those of a well-conditioned human athlete when tensed. I had the same impression of Nyota.

Nyota is a real showoff. He often pranced down the table with arms upraised like a hula dancer, and jumped into the air and came down with perfect balance. I imitated his dance (at least the part with the arms), and no doubt this semi-synchronized dance involving a 72-year-old man and an 18-month-old bonobo provided a rare sight. I'd love to have pictures.

After we had been inside 45 minutes or so, Duane, who had left to go shopping, returned with vegetables for the Japan trip, and I started feeling uncomfortable about taking so much time. Sue still was not completely ready for her lectures at Kyoto University, which made me feel extremely grateful and more than a little guilty. So I said it was time for us to go, and Sue picked up Nyota. I thoughtlessly came over to give both of them a simultaneous hug goodbye; Sue warned me to make sure that it was okay with Nyota, who was likely to be jealous about anyone touching his "mother" without permission. It seemed to be okay, but I was less effusive than I might otherwise have been. That marked the end of my great ape contacts for that trip to Atlanta, except for riding back to the airport with Duane.

AN OVERVIEW OF HER WORK BY
DR. SUE SAVAGE-RUMBAUGH

The research program at the center began in 1972 with the efforts of Dr. Duane Rumbaugh to determine whether it was possible for a chimpanzee to learn to recognize graphic symbols and to associate them with objects. The Gardners had already reported that Washoe was learning sign language. I had just begun graduate work at the University of Oklahoma, where Washoe had been relocated (with Dr. Roger Fouts) when she outgrew the house trailer in the Gardner's backyard in Reno, Nevada. My introduction to apes began when Dr. Fouts brought a 3-year-old chimpanzee named Booee to a psychology class and requested volunteers for a new program of sign language instruction that he was conducting at a chimpanzee colony outside of Norman, Oklahoma. At that time I was studying child development and had a 2-year-old son who was beginning to acquire language. I was amazed to see a chimpanzee sit in front of a class of 100 students and sign "hat," "key," "string," etc. as Dr. Fouts held these objects up one at a time. I volunteered to assist in his work.

As a result, I was introduced to a colony of 30 chimpanzees. Some of them lived in social groups in cages, others lived on a small island, and still others lived in human homes where they were being raised as human children as part of a research project directed by Dr. William Lemmon. His goal was to determine whether or not chimpanzees would exhibit sexual and social behaviors typical of chimpanzees if they were raised as social beings but without exposure to other chimpanzees. The differences in behavior between these groups of chimpanzees were striking, and I was fortunate enough to work with all three groups. Lucy, a chimpanzee reared in a human home, was the most surprising. She had learned many signs, and when I visited her in her home, she showed me how to make tea, how to wash the windows of the house, and one day even how to change the tire on my car. It became my duty to take Lucy, who was then 7 years old and weighed about 80 pounds, for rides in the car and walks in the woods as we worked on new signs.

The chimpanzees that were reared in social groups in cages could do none of the things that Lucy could. However, their social interactions were fascinating to observe, and I began to study the nonverbal communication that existed between mothers and infants. The juvenile chimpanzees on the island had all been reared by human beings for a few years and were then housed together. Lacking chimpanzee or human mothers, they were brought off the island each day. We volunteers worked to teach them signs and took them for walks. Thus I experienced chimpanzees reared by human parents with relative freedom, caged chimpanzees reared by caged chimpanzee mothers, and juvenile chimpanzees of various backgrounds who were trying to grow up with the help only of one another and who had some freedom from time to time. In addition, I was exposed daily to chimpanzees who had learned signs and those who had not. I also saw what happened to chimpanzees who were sent away from their human parents for various reasons and who were forced, late in life, to try to learn how to get along with other chimpanzees. Many years after this

I went to the Congo and saw wild bonobos living in kinship groups and raising their offspring in complete freedom. Then for the first time I saw the importance of the actions and cognizance of adult chimpanzees in a natural group making their way in the world. I saw firsthand how critical, communication, planning, and group co-ordination are in promoting daily survival.

I learned profound lessons from these experiences of the differential effects of socialization, language, and *Pan* versus *Homo* enculturation. These impressions shaped the course of my future work and in many ways shaped the questions that arose in my mind each time I encountered additional apes.

It was clear that juvenile chimpanzees raised without parents did not imitate our signs or learn language observationally, as do human children. They only learned signs when we employed Dr. Roger Fouts' technique of showing them an object and then molding their hands into the proper configuration. I puzzled over the meaning that these signs, so acquired, held for them. Clearly they knew which signs to make when shown an object, and they knew how to request things they wanted. But their language capacity appeared constrained in many regards. For one, our signs to them were not responded to as though they were attending to and comprehending information to be acknowledged and acted upon. Words seemed to lack an existence apart from the context of asking human beings to carry out simple actions such as tickling, transferring food, etc. Their language was a one-way goods-and-services affair. It was difficult to discern whether this limitation lay in them, the way in which they were being taught, their histories, or their current environment.

Unlike the juvenile chimpanzees, Washoe had been immersed in an environment in which she was signed to as well as taught to sign. However, her world at the University of Oklahoma differed greatly from what it had been when she was a child in Reno, Nevada. She was no longer surrounded daily by human companions and competent signers. Instead, she lived with chimpanzees, and, except for walks a few times each week with Dr. Fouts, she lacked opportunities to sign. Although she knew many signs, and watched Roger closely as he signed to her, her early enculturation had emphasized the capacity to produce signs; perhaps for this reason she seemed much less motivated than humans to respond to the signs of others by giving them food or objects, or by sharing novel information about events.

Were chimpanzees simply egotistical beings with little concern about the linguistic expressions of others? Or did a rearing environment which was focused upon a single ape, and encouraged that ape to express its every desire, result in an egotistical orientation to language? Although Washoe was treated in many ways as a human child, she was nevertheless a research subject. Of necessity, she had to be encouraged to develop some competency that could be evaluated within the paradigms of experimental psychology.

I wondered what my son's language would have been like had he been reared like Washoe. Would he have the freedom of expression that he did or would he too have a self-oriented set of words designed to obtain goods and services from those around him? What if the goal had been not to integrate him into the human community but to constantly question and evaluate his intellectual competency and to record his every utterance, often even before replying?

Later, Dr. Fouts moved to Central Washington University, where Washoe was joined by three other chimpanzees who had also been immersed in a signing environment by the Gardners. During the early years of their childhoods, each of these chimpanzees had replicated the early experiences of Washoe. Opportunities to use signs to communicate with other chimpanzees began to emerge. Their favorite signs became things that they requested from their caretakers. Though researchers would observe them signing to one another to request tickling, chasing, and food, their language would remain self-oriented and constrained mostly to the present. Since chimpanzees in the wild were quite capable of conveying such needs and desires without sign language, I wondered whether only a cultural difference in expression existed between them and captive apes. Was the language that was appearing reflecting the limits of the ape mind or the limits of captivity and training?

Following my graduation from the University of Oklahoma, I moved to Atlanta, where I encountered Lana and four juvenile chimpanzees that were learning to use lexigrams instead of signs as a means of communication. Lana had also been raised in an environment in which human companions communicated with her, but her world was limited to a single room with a large keyboard. While she could not go for walks, she could request that a computer make a window open, vend food, or show her slides when human beings were not present. When humans were present, Lana, like Washoe, asked them to do things for her. Lana, however, was taught to form complex requests such as PLEASE TIM PUT BANANA IN MACHINE or PLEASE TIM MOVE INTO ROOM. Her multi-symbol constructions contrasted sharply with utterances such as BANANA ME ME ME that characterized the apes who had learned to sign. Nonetheless, Lana's language seemed constrained as well. Like them, her ability to produce signs and symbols in response to human requests or experimentally generated situations greatly exceeded her capacity to comprehend symbols containing new information about the environment or to respond to the requests of others. She was raised first and foremost as an individual research subject. She was prodded and tested to acquire productive skills that would be impressive indications of intellectual competencies as assessed from the perspective of experimental psychology.

Thus my impression of the language capacities in apes formed from all of the apes that I had observed (both those who had learned symbols or signs and those who had not) was that what was lacking was (a) human-like appreciation for the communications of others, (b) an interest in communicating about events of the past or the future, (c) a social willingness to cooperatively coordinate anything other than immediate actions through language, and (d) a perspective on the function of language that was not completely self-oriented. These differences were understandable in that language training, regardless of its modality, had focused upon developing skills in producing language, rather than in comprehending and responding to language as employed by others. Moreover, each of the apes who had learned symbols had been treated as the focus of attention by a group of caretakers whose job was to increase the number of symbols that the ape could produce, and to respond to and record their every utterance. None of the apes were really incorporated into a social kinship community such as one might find in the wild, where the language and roles they were acquiring were designed to produce effective functioning adults who traveled in an integrated manner, protected themselves as a group, located their own resources based on their wits (and possibly upon their ability and willingness to share information about those resources with one another). In short, survival skills and group coordination were not a part of the language world to which the tutored apes were exposed. They had no real need to talk to each other about events in their environment, about the past and the future, or to coordinate their activities. Human caretakers whose main focus was to find ways to demonstrate the apes' intellects under controlled test situations provided these things. Thus the question remained: To what extent were we learning about the limitations of apes and to what extent were they merely reflecting the constraints of their environment and the limitations imposed by our failure to consider their communicative abilities within an adaptive context?

For this reason, when I moved to Yerkes, I began to focus upon what linguistic communication would look like if it occurred between chimpanzees, rather than between a chimpanzee and a human being who was always evaluating the correctness of what was said. Observations of social groups of wild chimpanzees suggested that they were adept at non-verbal communication and inference. Therefore it seemed that, if the languages they were learning really gave them cognitive skills beyond the non-linguistic ones they already possessed, these languages should enable inter-individual communication between chimpanzees in a new way. After all, that was what the invention of language had supposedly done for our species. It had set *Homo* on a new trajectory that enabled us to understand the world differently and to communicate with each other in profoundly new ways, ways that were said to lead to the cross-generational transmission of knowledge and to human culture. If language per se were that powerful, then it should ratchet chimpanzees to a new cognitive level.

Initially, I too assumed that chimpanzees did not possess language in the field. Therefore, the only way to determine what a chimpanzee who had language might be like was to teach it some language and then to determine the effect of one's teaching on aspects of its behavior. Thus I began by teaching lexigrams to four juvenile chimpanzees who were group reared and who, like those in Oklahoma, lacked mothers. Initially they were taken from their cages once or twice a day for training sessions, just as had been done at Oklahoma. Symbol acquisition was painfully slow. Learning how to ask for tickling, chasing, and food came easily, but learning to differentiate names of specific foods and specific objects did not. In fact, until a single caretaker was assigned to each young ape to serve as a parental surrogate who spent many hours each day with that young ape and became their main social companion and teacher, there were no breakthroughs. I was reminded strongly of Helen Keller's account of her entry into language, which required the establishment of a similar relationship.

Once the young apes had learned to associate symbols and objects, I asked them to do three things not typically required of other apes that had acquired language. I asked them (a) to listen to me and respond behaviorally to my requests and needs, (b) to listen to each other's requests and needs for specific items beyond things like tickling, chasing, and food sharing that could be conveyed non-verbally, and (c) to make statements about future intended actions. Even though these young apes were proficient with many symbols, they could not do these things with words.

Their failure opened a new window upon what the world of language was all about, how it operated, what it does for us, and how it does it. And I began to understand why it was that the language of the apes had seemed so constrained. The essence of language lay in the social contract that formed between speaker and listener within the bounds of a cultural community. Words acted merely as the manifestation of this multi-layered contract, and the teaching of words without enabling the social contract produced the constrained language that so puzzled me. I saw that in the normal rearing of a human child, the social contract begins at birth, with the first cry, the first touch, the first eye contact, and the way in which each of these things is patterned into language. I realized how much of language is understood, long before the first words are spoken. The central duty of the human infant is to learn to participate in the social contract and to coordinate its behavior with others. Language emerges as a vehicle for that coordination. Consequently, listening and coordinating, not speaking, become the primary vehicles for acquiring language and for enculturating oneself into the larger community. In their desire to develop language in apes, researchers focused upon what apes could produce rather than upon what they could understand, and they over-looked the fundamental interaction between socialization and language that comes about first through a primary caretaker and later through the community at large. The human infant's immersion in a language environment requires the negotiation of social contracts in ways that entail respect for, cooperation with, and comprehension of the general intentionality and specific intentions of others. These come about with the unstated goal of producing a socially competent adult who will function as a member of a specific society. The experimenters who were rearing apes did not set out with this goal. Just as they believed that language had to be taught, they also believed that apes, as they grew past childhood, would not turn into competent adults within a human society. They would, at best, be retarded and estranged beings needing restraint and human care, or they would have to be placed with other captive apes to live in cages, in zoos, or in research centers.

With these realizations, my training strategies moved away from words and began to focus on social contracts. The young apes learned to share food with each other, to attend to each other's requests, to comply with them, to state their actions in advance of simply acting so that others could elect to coordinate, or to disagree. Life began to center around routines, activities, and plans for the day, shared and conveyed through symbols. How to obtain the food was planned, how to store it was part of the routine, how to prepare it was planned, helping to prepare it was planned, outings and different activities each day were planned, and innovation in the planning was encouraged. Out of this stuff of life a language that demonstrated behavioral coordination and reciprocity began to emerge. Statements of goals appeared, negotiation appeared, self-monitoring and correction of errors appeared, putting away toys and keeping areas neat appeared, concern about how food was mixed and cooked appeared, self-recognition in the TV monitor appeared, differentiation of past and present representations of self appeared. But three things were still missing. Comprehension of novel spoken sentences, apart from context, did not emerge. Syntax did not emerge. And the moral ethics that are formed by enculturation into the larger community that exists beyond the home (or laboratory) did not emerge, as the apes were not permitted access to this community.

Washoe, Lana, and the other apes mentioned above were of course not the only subjects of scientific experimentation. In the 1950s the Hayeses reared Viki as a human child to determine whether she would learn to speak. She did not do so spontaneously, and they resorted to training as well. Even with training, speech proved difficult, and Viki learned to produce only a few humanlike sounds. They concluded that speech was nearly impossible for chimpanzees.

Dr. Terrace set out to rear a young chimpanzee named Nim in an environment of sign language immersion, and he too found training necessary. Terrace decided not only that training was essential, but also that the kind of language chimpanzees were able to produce differed profoundly from that of children. While Nim learned words, Terrace found that he tended to rely on imitation when asked to converse. Terrace believed that Nim's understanding of language had not traveled beyond associationistic connections between word and sign. According to Terrace, this was because Nim was unable to comprehend or to produce syntax. Syntax, according to many linguists, is seen as the structural feature of language that enables words within a string to reference and define each other in certain ways. Formal linguists proclaim syntax as the quintessential component of human language. It is said to permit us to move from an associationistic linking of referent and object, to one that frees symbols for virtually unlimited forms of meaning, including metaphor. Syntax provides the underlying algorithms of language and thought. These algorithms allow the human brain to free itself from the immediate present. Such algorithms, in the dawning of the computer age of artificial intelligence, exert a strong appeal. Everyone understands that computers cannot function without them. Moreover, these algorithms are structural in nature (if-then; and, first, second, third, etc.), and they are information independent. That is to say, they operate in the same manner regardless of the information that is placed upon the structure of the program itself. Until the appearance of neural nets, without such algorithms computer programs could not be built. This led to what seemed to be an obvious conclusion, namely that the chimpanzee brain lacked the algorithms for the structural components of language, and thus a true language could not be constructed with the nervous system of an ape. By contrast, the human brain, for reasons that were hidden deep in evolutionary history, had achieved an instinctive capacity for such algorithms, and it needed only experiential input, with any language, to awaken them and put them to work building an internal linguistic program. Such views were, and are, held, independently of any data that reveal the capacity of apes to engage in syntactical constructions.

In Japan, Dr. Matsuzawa began work with a chimpanzee named Ai, who was taught to use lexical symbols. Ai's environment was not designed to foster communication or syntax. It focused instead upon assessing chimpanzee intelligence independently of characteristics cherished as innate by the genus *Homo*. Thus Ai learned to associate symbols with objects, colors, numbers, etc., but not to employ her symbols to request goods and services from human beings, to engage in conversations or to listen in any way to language directed to her. She demonstrated incredible feats of short-term memory, the capacity to count to nine, the ability to copy symbols, the ability to learn and to sequence colors, names and quantities of objects, and the ability to outperform human beings in many tasks requiring intellectual speed of recognition. This work has gained respect for the cleverness of chimpanzees, while avoiding the contentious issues of meaning, intentionality, and syntax.

Dr. Patterson began research with a gorilla named Koko by assuming that language training was essential. Initially her work took place in the zoo, but Koko was soon moved to a house trailer where immersion in a signing environment took place. Koko was exposed not only to signs, but also to spoken English in combination with sign. Today she is said to know more signs than any other living nonhuman primate. Like the apes described above, Koko's caretakers have responded to her every utterance, and their focus has been upon teaching her as many signs as possible. Koko is reported to answer questions about the past and the future and to be aware of the nature of death. Penny also began teaching signs to a male gorilla named Michael, who was reported to have been able to describe the traumatizing conditions of his capture in the wild. Many scientists view these results with skepticism because they think that Patterson offers insufficient proof.

However, the difficulty in offering proof of such elusive statements makes us ask what is acceptable as proof. Statements based on memory in the human species are taken at face value and are accepted in court. If they do not agree with the memories of others, we may assume that one party is lying or mistaken, but we nonetheless assume that both parties are capable of reporting upon past events (either correctly or incorrectly). This assumption is based upon our own subjective knowledge that we can do so. Since we are not gorillas, we hesitate to use our subjective knowledge to extend that assumption to gorillas. Lacking subjective extension, we demand proof. But proof of the subjective content of memories or thoughts about death, childhood trauma, and other non-verifiable events can only be obtained by self-report among human beings. If we deny the validity of self-report in other species, we create conditions for proof that are unachievable.

The studies above are described for the purpose of putting the current and future work with Kanzi, Panbanisha, and Panbanisha's offspring in proper perspective. Kanzi was the first to help us realize that the capacities of apes were far greater than we had thought and that the emergence of their skills was not dependent upon teaching but rather upon the building of a proper physical and social environment. Kanzi continues to help us realize this in new ways. This work with bonobos shares some attributes with the studies above, but it differs in fundamental ways as well. Its findings and its goals for the future cannot be described without properly explicating both the similarities and the differences.

The significant differences between the bonobos at LRC and other apes are:

(1) Kanzi and Panbanisha were not reared in the absence of a bonobo mother. Matata has been a part of their world since their birth, as have many of her other children, who are their siblings. Thus Kanzi and Panbanisha's world was, during their early years, a bi-cultural one, composed of human beings and bonobos. These people interacted with Matata as well as with her offspring. Kanzi, Matata, and human beings moved together through a forest searching for food as a socio-cultural group. Later Kanzi and Panbanisha continued to move through this environment, while Matata stayed behind and raised more offspring that did not share this bicultural experience.

(2) In their world Kanzi and Panbanisha heard vocalizations from Matata and their human caretakers, and they watched lexical symbols being used to communicate coordinated actions and goals.

(3) Their comprehension and acquisition of bonobo vocalizations from Matata was spontaneous in that she did not train or reward them for so doing. She did, however, serve as an important model, reliably producing important vocal information for them on appropriate occasions. Similarly, their comprehension of human spoken and lexical utterances was spontaneous and occurred without reward or training. Their human companions, like Matata, served as important models for their communicative capacities. Many times, the kind of things that they communicated differed from those that Matata, who spoke far less, elected to convey.

(4) Kanzi's and Panbanisha's English comprehension clearly encompassed the structural and syntactical nature of language. They understood novel sentences and complex narrations when they were 5 years old. However, their production of English was compromised by their anatomy.

(5) The productive use of printed lexigrams was spontaneous and they, like Koko, communicated about the past, the present, and the future, as well as about the intentions and thoughts of others. In many cases we have been able to verify these communications. We have not yet attempted to probe their understanding of death; however, a number of individuals known to them have died, and it is possible to go beyond self-report in probing these sensitive issues because of their great command of English comprehension and their willingness to answer questions about their thoughts and about past events.

(6) They have begun to translate for Matata and for their other siblings. Verified translations include the following:
   (a) Matata had seen a stranger in a tree (which turned out to be a telephone pole man)
   (b) Foods that she wanted to eat
   (c) Foods that had been brought into her building
   (d) Visitors who had arrived
   (e) Doors that she wanted open
   (f) Requests that she made to visit certain areas
   (g) Actions that she had done to them (such as biting)

Kanzi and Panbanisha, having both reached high levels of English comprehension, have begun to employ a bonobo Creole of English with each other.

At times, they are able to translate for us the meanings of these sounds. As such they are the only bi-cultural apes that are attempting a spoken language, that comprehend the syntax of English and that listen to and respond to language in ways that are free from the constraints of training paradigms. This is not to say that they are more intelligent than other apes; however, it is to declare that their bicultural competencies exceed those of other apes, that their skills of translation between English and bonobo vocalizations are unique, and that the flexibility of their linguistic capacities is unusually great.

They have demonstrated clear comprehension of the key structural algorithms of syntax (such as word order, recursion, possession, agency, and metaphor). They have spontaneously begun to make rock tools after observing human companions do so. They have spontaneously begun to make attempts to write lexical symbols and to draw. They have shown some spontaneous capacity for music, mythology, and narrative.

The essential distinction between Kanzi and Panbanisha and others noted above is that their capacities have not required tutoring because they emerged spontaneously in a bicultural world of forest travel and freedom to do as they pleased. Moreover, they shared these experiences with other bonobos, not solely with human beings. Their language is not constrained to demands for goods and services. They request things for others, report past events, plan actions for the future, tell us who should be taking care of them, remind us of things we have forgotten, announce events that are occurring at the lab and, if asked, will comment upon their internal state and their thoughts.

Thus they have developed a bi-species culture, small though it may be. Such a culture is capable of propelling itself forward, in time, of its own accord under the proper conditions. Like all cultures, it is able to construct new ways of doing and being from the pieces of the old. It is not that this makes them better than other apes, or more intelligent. However, it does make them distinctly different. They can translate things between cultures, and they have offspring to whom many of these skills can be transmitted.

SAVAGE-RUMBAUGH'S WORK: A VIEW FROM OUTSIDE

Davis and Balfour in their book, *The inevitable bond,*[2] argued that people who work with animals cannot avoid empathizing with their subjects. My experience at the LRC gave me a vivid demonstration of the correctness of that view, at least when the animals in question are young bonobos. Only a person with animal phobias, or lacking in the ability to form attachments, could fail to feel close to such a subject. Critics of animal language research in particular may bemoan the anthropomorphism that could accompany this sympathy for the subjects. And admittedly, objectivity is generally accepted as a hallmark of scientific research.

There is, however, another view that may be equally legitimate. The meanings of behavior are not always clear unless the context of the behavior is taken into consideration. That context may be more accurately observed and evaluated by scientists who understand their subjects from the inside out. They may be better able to enculturate and teach their subjects, as well as to understand them better. It is also tremendously encouraging from the viewpoint of the welfare of animals and humans alike that almost every scientist who has been involved in long-term animal language research has also become very involved in efforts to conserve animal populations in either captive or natural settings— or more often in both. So I applaud the formation of the inevitable bond on balance; its scientific merits probably outweigh its costs, and its nonscientific merits are unquestionable. I am proud to be among the tree (and bonobo) huggers.

A SKEPTICAL VIEW

Dr. Sue Savage-Rumbaugh is no soft-headed woman, despite her dedication to chimpanzees and bonobos of all ages. In the beginning, when she started work in Oklahoma, she did not believe that other researchers had demonstrated that chimpanzees truly understood words. She knew that a rigorous demonstration was far more difficult than most people assumed. It required that systematic research show that lexigram symbols are *really* grounded for the chimpanzee, that they stand for the thing to which, for humans, they refer.

She started work with four animals, Erika, Kenton, Sherman, and Austin. However, the demands of working intensively with four animals proved too great, and only Sherman and Austin completed the 8 years of training that preceded the writing of her book, *Ape language: From conditioned response to symbol.*[3] The title reflected her belief that the ape must do more with the lexigram, hand movement, or piece of plastic than simply use it to obtain food. True symbolic use required that the ape understand that the symbols he or she used *stood for* something.

It is easy to assume that animals understand what we say to them; our assumption seems supported by their ability to execute what we tell them to do. However, fairly early in her work with Sherman and Austin, Dr. Savage-Rumbaugh showed that they, despite their following of commands, understood little or no spoken English. If contextual cues were eliminated, they were unable to follow commands or pick out objects requested in English. Thus a casual observation that an organism responds to commands like "stay" or "heel" tells us nothing about whether the organism understands the words as words.

Methods of Proof

As part of her preparation for teaching Sherman and Austin, Dr. Savage-Rumbaugh made a careful analysis of how human children learn language. She based her analysis on the findings of Elizabeth Bates:[3] "Bates (1976) found that during the development of preverbal communicative competence, the human child passes through three stages: 'from exhibiting self, to showing objects, to giving and pointing to objects'" (p. 11; cf. 4, p. 61). Chimpanzees also show off themselves and objects, but do not give anything willingly and do not point.

Dr. Savage-Rumbaugh wanted to teach the chimpanzees to label objects, and chimpanzees naturally show the most interest in objects that are edible. However, she found that showing an object and requiring the chimp to press a matching key to obtain a single reward (a slice of apple, for example) was incredibly unsuccessful. This is a "single reward" condition because the same reward is given, regardless of the object being matched. Yet when there were several keys, different key presses delivered different rewards associated with the individual key, the animals learned relatively well. They demonstrated their learning by requesting preferred foods first nearly all the time. This "differential outcome" procedure was much better than the "single outcome" condition, which seemed to make no sense to the chimpanzees.

Learning to Share

After many trials, errors, and successes, Sherman and Austin learned to request and name a number of foods. However, in order to demonstrate true communication between animal subjects, Dr. Savage-Rumbaugh needed to design a situation in which the chimpanzees had to use symbols to solve a problem whose solution would benefit both of them. The obvious solution was to require the use of symbols to obtain food that both animals would share. The problem was that chimpanzees, like young humans, are very reluctant to share—in fact, they almost never do so willingly in their natural environments, the exceptions being occasional sharing between mothers and infants or sharing small portions of meat after a kill.

Sherman and Austin were, therefore, laboriously taught to share and then taught to request certain foods from each other in the presence of a cafeteria of foods. Doing so initially produced some hilarious (in retrospect) episodes, as when Sherman became quite upset when the experimenter gave Austin the first share, or when Austin, the submissive chimpanzee, was extremely reluctant to take food from Sherman even when Sherman offered it. However, they did learn to share, and they then requested and shared even when the experimenter was not present, apparently enjoying this game, which they played with considerable accuracy (Sherman apparently better than Austin). Sharing became a culturally expected process. They also learned to request tools from each other (wrenches, keys, etc.,) so that they could gain access to locked food boxes.

Animal-to-Animal Communication

Sherman and Austin also learned to attend to each other in a communicative setting, which they did not do at first. They did better at this when put into separate rooms and required to communicate via keyboard. Later, however, they were put together in a single room with an array of foods. Sherman, Austin, and Lana all learned to request specific foods and label foods without eating them, and also to label objects. Only Sherman and Austin learned to request and give objects and foods. Lana learned to select strings of lexigrams. Sherman and Austin also learned to label food and tool objects according to function—even unique foods or tools. Lana was unable to label new foods or tools; she never got the idea in this sense, but *she was able to sort the same objects with 100 percent accuracy.* Sherman could also label photographs correctly, but Austin labeled at about chance.

Classifying Lexigrams

As a final proof that they understood the meanings of symbols, Sherman and Austin both labeled, with very high accuracy, lexigrams that they had never seen labeled before as food or tool. This was a highly abstract task that could not have been done on the basis of the physical similarity of the food lexigram to the lexigrams for the specific foods or of the tool lexigrams to the tools. One error that Sherman made, labeling the lexigram for sponge as a food rather than as a tool, was easily comprehended because Sherman often ate the sponge (the sponge was used as a tool to soak up drinkable items like juices or coke). Dr. Savage-Rumbaugh comments:[3]

> Lana's performance, as contrasted with that of Sherman and Austin, revealed clearly that different chimpanzees who learn symbols may not have learned the same thing. Training variables are exceedingly important. It is necessary to question the degree to which vocabulary items really stand for things as used by language-trained apes, particularly in projects where no explicit effort has been made either to teach or test comprehension (p. 265).

Spontaneous Utterances

Lexigrams were used spontaneously as well as in formal training sessions. Most such utterances were one to three symbols in length, but some were longer. Many combinations were imitations of what an experimenter had said, but others were partial imitations and still others (about 25 percent overall) were completely unique. Sherman and Austin typically combined gestures with lexigrams to clarify requests. Their lexigram utterances were usually considered correct in the context. In one charming incident, Sherman combined the COLLAR lexigram (indicating that the experimenter should put the collar on him and take him out) with a picture of Koko to ask to go see the gorillas near his quarters. When he got close, the trainer said NO because Sherman upset the larger animals, negating any ongoing research. Sherman discarded the magazine containing Koko's picture and looked very disappointed.

Apes, Pictures, and TV

Most apes prefer still pictures in magazines to moving pictures. Washoe, Viki, Lucy, and Koko all showed this preference. It is surprising that Premack and Woodruff[5] were willing to assume that problems could be presented to their star chimpanzee Sarah via TV. A problem might consist of a trainer who pretended to try to get water out of a hose that wasn't connected, and Sarah was to choose the "solution" from two 35 mm photographs, one of which showed the hose properly connected. It is not clear why Sarah should have interpreted the presentations as problems for the trainer, to be solved by Sarah. Sarah could have chosen the correct solutions for other reasons, for example, that the solution was a better physical match to the problem. However, Sarah's production of a high percentage of correct matches does prove that she attended either to some aspect of the 30-second video clip that preceded her response, or preferred the correct photograph on other grounds.

As a comparison, a similar experiment was performed with Sherman and Austin. In this experiment, no problem was presented at all, but they chose the same "solutions" on 25 of 28 trials. They apparently shared preferences with each other and with Sarah.

Sherman and Austin, like the other apes, initially were not interested in TV. Then they were trained with larger, better, pictures, and apes they knew were the actors. They got more and more interested until they were watching for 30 minutes at a time, and making behavioral responses like displaying, approaching, and biting the screen. They then transferred their interest to a small video screen, only 8 inches diagonally, and could label food, etc., shown on it. The experimenters then showed food in an adjacent room, and Sherman and Austin, catching on quickly, asked for it, brought it back to each other when told to do so, etc. They recognized themselves on TV, too. Austin did so quickly, Sherman only after months of failure. In one striking photograph, an open-mouthed Austin is shown shining a light down his throat while exposing it to a video camera and watching the picture on a television monitor; he had obtained the light in order to get a better view of his usually dark throat!

The experimenters also used the TV to set up food exchanges, showing the available foods on the monitor. Austin and Sherman did well on this task, requesting mostly only the available foods and generally retrieving the food requested. They also quickly learned to operate joysticks after observing people controlling an arrow via a joystick. They were good at getting the arrow to the right place. They had apparently acquired a broader conception of possible causal relationships. Non-language-trained chimps showed not a hint that they understood the joystick situation.

Studying Spontaneity

Sherman and Austin, as stated above, made spontaneous statements, often labeling something or announcing what object they were about to give to the experimenter in a test situation. However, their most frequent use of language was for requesting. Lucy, the Temerlins' signing chimpanzee, apparently made more spontaneous statements with less training, as allegedly did signing chimpanzees like Washoe, or the gorilla Koko. However, Lucy did not seem to be *commenting to* anyone, just emitting the signs. She could not retrieve objects that were not present, or pick the correct one out of several, or make a statement about what she was about to do. Her signs were like conditioned responses, tied to the here and now. Thus Sherman and Austin went well beyond Lana and Lucy in their ability to communicate meaningfully with each other and with their trainers, and the care with which the experimenters supported this claim was unprecedented.

Thorough observations over several sessions indicated that few of Sherman and Austin's utterances (unlike Nim's) were complete imitations, and they made many spontaneous utterances. It appears that the use of a lexigram board is less likely to elicit imitation than is use of ASL. It is also likely that the keyboard encourages turn-taking rather than the interruptions so often seen in Nim. The multi-lexigram utterances of Sherman and Austin were generally reasonable and designed to transmit information. They did not appear to be random selections of symbols, although their symbol strings manifested no detectable formal grammar. Among the things that Sherman and Austin could do through their use of lexigrams and gestures were coordinating activities, making comments, requesting, referring to things not present, announcing action, and learning from one another.

APPLICATION TO HUMAN CHILDREN

The results with Lana, Sherman, and Austin encouraged the Rumbaughs and their coworkers to apply the lexigram technique to children who had little or no ability to use vocal language. Note that lexigrams allow symbols to be recognized rather than recalled, as they must be for vocal or sign language to be used, thus easing the memory burden on slow learners. Romski and Savage-Rumbaugh[6] described 14 children, none of whom possessed more than minimal language skills, with whom they worked. They had some success with eight of them, all of whom might fairly have been described as hopeless because of their failure to learn language through other modalities and techniques. Two of the fourteen children left the institution, two were too aggressive to handle, and two failed.

Lana's success with her lexigram board encouraged researchers to develop and test a similar portable lexigram apparatus that could be used with severely language-handicapped children, who in most cases achieved some ability to communicate, when previously they had had none or nearly none. Photo courtesy Dr. Duane Rumbaugh.

Kanzi (the name means bold and brave in Swahili) has demonstrated more understanding of English in careful tests than any other nonhuman animal, as well as a well-developed ability to knap stone to make stone knives.

THE STAR PUPILS: KANZI AND PANBANISHA

Bonobos and Chimpanzees

In the early 1900s there was a growing awareness in the world of western science that there might be two species of chimpanzee-like creatures, rather than just the one previously recognized by the scientific community. The "common" chimpanzee, *Pan troglodytes*, was the only species recognized. However, in 1916 the Dutch naturalist Anton Portielje suggested that a very popular animal named Mafuca in the Amsterdam zoo might be a separate species, and a few years later the American psychologist, Robert Yerkes, made the same suggestion about one of his animals, Prince Chim.[7] Little attention was paid to these suggestions, however, until two anatomists, the German Ernst Schwarz and the American Harold Coolidge, made the distinction on the basis of a skull they observed in a Belgian museum. Schwarz published first, but Coolidge claimed in 1984 that he had made the distinction first and told the museum director at Teveuren, who passed the news along to Schwarz. Schwarz published in the *Revue Zoologique* of the Congo Museum before he, Coolidge, could publish.

Coolidge may, therefore, have been scooped by Schwarz. Both were probably scooped by natives who lived in the vicinity of the animals, but were short on scientific publications in which they could document their knowledge. In any case, the bonobo, *Pan paniscus*, is now recognized as a species separate from the chimpanzee.

Yerkes regarded his bonobo, Prince Chim, as a chimpanzee genius, and it is possible that Richard Lynch Garner's favorite animal, Aaron, was also a bonobo; Garner described Aaron as being intelligent and having an expressive face, two of several characteristics that are more common and humanlike in bonobos than in chimpanzees. Another of Garner's apes, Elisheba, was said to be a kulu-kamba, a type of ape whose status is still uncertain. Suggestions are that it is a cross between bonobos and chimpanzees or a separate species. Recent studies suggest that the anatomical features of alleged kulu-kambas fall within the range of variation of *Pan troglodytes*, so it is possible that they are not separate after all. It is an interesting puzzle.

Kanzi: Sue Savage-Rumbaugh has worked with several bonobos, two of whom, Kanzi and Panbanisha, probably comprehend English better than any other apes observed up to the present. Possible exceptions are the gorilla Koko and the orangutan, Chantek, whose talents in comprehending English have not been documented as carefully as those of the bonobos. Both Kanzi and Panbanisha learned largely by observation and natural interaction with lexigram boards and experimenters. Thus their learning is unlike that of most other animals, who were sometimes subjected to operant techniques that involved pairing symbols and objects and reinforcing correct responses.

Some of the better leads in scientific inquiry are serendipitous. So it was when Kanzi, at age $2\frac{1}{2}$ years, showed Dr. Savage-Rumbaugh and her research team that he could understand and use some word lexigrams. Kanzi, whose name in Swahili means "bold and brave," was born at the Yerkes Regional Primate Research Center on October 28, 1980. His birth mother, Lorel, was on loan from the San Diego Zoo, whose administrators decided that she should give birth in the colony. Lorel promptly lost Kanzi to the higher-ranking Matata, who already had an infant and thus was able to nurse Kanzi, as well as her own infant. That arrangement was fine with Kanzi and Matata, and Lorel put up only minimal resistance.

Kanzi started learning about language when he was about 6 months old. At that time Dr. Savage-Rumbaugh resumed an interrupted attempt to teach Kanzi's stepmother, Matata, to use the symbols on a lexigram board. Matata had been captured in the wild, and did not begin her lexigram training until she was nearly an adult, a year before the birth of her first infant, Akili. She had been a poor pupil before her training was interrupted, and her later performance was equally disappointing; after 2 years and 30,000 trials, she could make requests with only six lexigrams, and could not use them to name the associated objects.

The methods of instruction used with her were those proven to be effective with Sherman and Austin. With Matata, however, they did not work as well. What she learned one day she usually forgot overnight. Matata had been brought to the United States with four others by another institution to be used in biomedical research. That research was never conducted. Thus, it was only at an estimated age of 6 to 7 years that Matata was assigned to language research.

Despite the early disappointment with Matata, Dr. Savage-Rumbaugh was optimistic as she restarted her training. Matata had daily language instruction. Kanzi was present because the researchers had no choice. He was less than a year old when Matata's second bout of instruction began. Infant apes protest violently when involuntarily separated from their mothers, and their mothers, in turn, do all they can to prevent or reverse involuntary separations. Although

Matata had stolen Kanzi from another bonobo, she was Kanzi's mother so far as Kanzi and Matata were concerned. So Kanzi attended the language instruction sessions, but was never intentionally taught by his caregivers. During the sessions, Kanzi was a great nuisance. He would grab objects from whomever, steal food from Matata, jump from above on Matata's head and into her lap, and so on. Still, he did not deny Matata all chance to learn.

Although he was not included in the training, being regarded as too young, he was sometimes fascinated by the lexigrams, and would try to catch them as they lit up when their keys were pressed. By the time he was 14 months old, he occasionally pressed keys and then ran to the vending machine, having apparently made the connection between key presses and the delivery of food. At this time, however, he gave no sign of understanding that a specific key was related to a particular food. When he was 24 months old, a favorite key became CHASE; he would press it and look over his shoulder for the result, hoping that the experimenter would agree to chase him.

Kanzi's learning to that point had been completely spontaneous. He had received no training, but had simply observed Matata's training. Nevertheless, he not only learned something about lexigrams, but also invented iconic gestures, for example, extending his arm in the direction of a desired direction of travel, and making a twisting motion with his hands to indicate that he wanted a jar opened. He also indicated that he wanted a person to give him an object by gesturing first at the object, and then at the person, much as a human child does.

When Kanzi was about $2\frac{1}{2}$ years old, Matata was removed from the Language Research Center where she and Kanzi had been training. It was time for her to breed again, and she was taken back to Yerkes to mate with Bosonjo, a male who had accompanied her from the Congo. Kanzi was old enough to stay at the lab by himself, and he was lured into another area so that Matata could be taken away.

Thereupon, Kanzi's own training began. The experimenters were shocked when Kanzi used the lexigram board 120 times on the first day, for the most part correctly! He not only used lexigrams like CHASE and APPLE correctly, but also pressed other lexigrams that stood for other foods and then selected the associated foods when allowed access to the refrigerator. That was a strong suggestion that he was announcing an imminent action, something that Matata had never done and that Sherman and Austin did only after years of intensive training.

It was difficult to believe that Kanzi had picked up all these skills just by observing Matata's training, so the experimenters wanted to test his ability to be sure that he was not somehow fooling them. Testing initially proved impossible; Kanzi would only do what he was interested in doing. So the experimenters had to devise techniques that more naturally and casually revealed what Kanzi knew. His learning by observation also convinced them that they should abandon traditional training methods and allow Kanzi to acquire language as humans do, in an environment rich in opportunities to observe and use language. He would learn by living and by observing others' use of word-lexigrams. Thus, Kanzi was allowed simply to continue to grow up, much as a human child would in a family setting.

And so he did. Because bonobos and chimpanzees spend so much time foraging for food, food sites were established in 17 places throughout the 50 acres of forest surrounding the LRC. Kanzi could use lexigrams to indicate where he wanted to go in the forest, and in all of his interactions with experimenters back in the buildings of the LRC. Experimenters simply acted as though Kanzi always meant what he "said." When he chose a lexigram, they gave him the associated food or object, or engaged in the requested behavior.

This procedure was intended to teach Kanzi that the word-lexigrams named things or announced imminent events involving play or travel, and so on. Kanzi could therefore learn for himself what the experimenters intended each lexigram to mean. Thus, if he pressed A-FRAME, one of 17 sites in the 50 acres surrounding the laboratory, people would say "Oh, you want to go to A-frame." They would then pack a few things and carry a picture of the A-framed structure, their destination. That Kanzi benefitted from this kind of activity was indicated when he started to trot down the trail that led to the A-frame and, at junctures in the interlacing pattern of trails through the forest, chose the correct path. Kanzi was learning the forest. Eventually, he was able to take a person who was totally ignorant about the sites in the forest to any site that he/she specified. He even suggested shortcuts at times, but his suggestions were generally declined because of brush, mud, and snakes. Kanzi could count on getting favorite foods, different at each site, and provisioned fresh daily. His favorite? M & Ms.

Within 4 months Kanzi's lexigram vocabulary rose from 8 to 20, and he could find his way to all 17 food sites. Within 17 months, his lexigram vocabulary had increased to about 50, and he had started to use combinations of words shortly after his separation from Matata. Most combinations consisted of only two words, but some consisted of three. In many combinations, gestures were substituted for lexigrams, and counted as words; for example, Kanzi often indicated the actor or recipient of an action by pointing rather than by naming on the lexigram board. Kanzi's rate of symbol emission was not high, but his accuracy was good, most strings were spontaneous, and repetition within strings was not excessive.

After another year had passed, his caregivers reported that Kanzi seemed to comprehend some English words, more with each passing week. The first indication was that if anyone were asked to turn the light on or off, Kanzi would run over, jump up on the counter, and flip the switch. Sometimes he would toggle the switch back and forth and watch the light go on and off. It seemed clear that he had induced the cause–effect relationship between the position of the switch and the state of the lights in the room.

Later, Kanzi readily picked appropriate objects specified by English words, and also chose the lexigrams that stood for the words. Sherman and Austin could associate objects with lexigrams, but showed no comprehension of English under controlled conditions, despite giving the impression that they understood. (They may have understood some English when they could see the speaker's face, as well as hear the words; that possibility was never rigorously tested.) Kanzi was the first ape to show under carefully controlled conditions that he comprehended a substantial amount of English when it was presented in unique sentences that he had never heard before.

The most impressive demonstration of Kanzi's comprehension came in a test comparing his ability to that of a 2-year-old child, Alia, the daughter of Jeannine Murphy, who worked at the LRC. By the time of this test, it was clear that the number of spoken English words that Kanzi comprehended far exceeded the number of lexigrams that he used, although the latter number was also quite impressive. Kanzi was a little over $7\frac{1}{2}$ years old when the tests began.

In order to exclude the possibility that Kanzi or Alia might previously have learned to make rote, memorized responses to spoken commands, 660 novel sentences were used; that is, neither Alia nor Kanzi had heard the sentences before, and so could not have been trained to respond correctly to them. Rather, in order to respond correctly, they had to understand the sentences by analyzing their meanings. The person reading the sentences was behind a one-way mirror and could not be seen by Alia or Kanzi. A second experimenter in the room with Kanzi or Alia was prevented from hearing the sentences by wearing headphones playing loud music, and hence could not cue Kanzi or Alia to make the correct response. Many of the sentences were not only novel, but also unusual; one example was a sentence instructing them to wash hot dogs[9] (p. 271).

Testing of Kanzi and Alia continued for 9 months, at the end of which Kanzi was over 8 and Alia approaching 3. Kanzi responded correctly to 74 percent of the sentences, and Alia responded correctly to 65 percent. Kanzi had the most trouble when asked to operate on two objects, as in "Give Sue the hat and the potato," whereupon he usually gave her one or the other, but not both. However, if told to "Put the hat on the potato," he did so with no difficulty. Kanzi also did well with embedded phrases; for example, he did better with "Get the ball that's in the group room" than he did with "Go to the group room and get the ball." Amazingly, when given a command like "Kanzi, can you give the doggie a shot," Kanzi explored the objects available, picked up the toy doggie, found a syringe, uncapped it, stuck the needle in the doggie, and pushed on the plunger. This analysis of a fairly complex novel sentence, as indicated by a completely correct response, went far beyond what had previously been demonstrated under carefully controlled conditions with any nonhuman animal.

This is not to say that Kanzi was perfect; 74 percent correct is not 100 percent correct, and far more complex commands could, of course, be designed. Some of Kanzi's errors were clearly misunderstandings of words; for example, when told repeatedly to put paint in the potty, he repeatedly put clay in instead, and finally, in apparent disgust, brought the potty full of clay to the experimenter to show her that he had done what she told him to do (which, of course, he had not, but he was making the reason for his error abundantly clear). Despite his imperfection, Kanzi's performance was astonishing, even earth-shaking; his comprehension of these novel sentences gives us an insight into the bonobo mind that was never before available. He apparently was able to segment the stream of vocal language, find the words, analyze them, understand the structure of the command, and carry it out despite its uniqueness. His performance was more than impressive; it was mind-boggling.

Kanzi's vocalizations were, nevertheless, quite limited. A study of the vocalizations of five bonobos revealed 14 distinguishable sounds, 10 of which were common to all of the animals. Four were unique to Kanzi, indicating that his language exposure had increased even his vocal vocabulary to some extent. In addition, sometimes Kanzi's vocalizations had a multisyllabic structure, with distinguishable units, unlike the typical bonobo vocalization with its single-syllable structure. Kanzi sometimes seems to be trying to imitate human vocalizations. Why he and other apes have such difficulty with producing vocal language remains a mystery, despite considerable study and even more speculation. We have described some of the hypotheses earlier in this book. Whatever the reason or reasons, it is always too early to say that they will never be able to talk! In fact, as she wrote earlier in this chapter, Dr. Savage-Rumbaugh believes that Kanzi and Panbanisha are starting to use a Creole of bonobo and English.

A recent paper[10] provides support for the idea that bonobos produce distinct vocalizations in the presence of different foods or in saying "yes." Vocalizations in the context of grapes, bananas, juice, and asking a question to which the expected answer was "yes" differed on several acoustic measures, including duration, onset to peak time, peak-to-minimum range, peak-to-endpoint range, and the slope of amplitude from peak to endpoint. Thus the subjective belief that bonobos transmit information via the vocal modality is supported by physical measures.

Sally Boysen has formed the same impression by listening to her chimpanzees. On a recent Discovery channel program she reported that one day she was sitting in her office listening to chimpanzee vocalizations while an assistant was feeding them. She thought "they must be getting grapes today." Upon checking, indeed they were. This led her to believe, a belief later verified, that if she could understand the vocalizations the chimpanzees could understand them too!

Even if the nervous systems or vocal tracts of all apes proved to be incapable of generating human vocal language, it might nevertheless be possible for apes to communicate better vocally by means of a prosthesis. We have mentioned that David Premack designed a device that used the position of a joystick to specify which phoneme would be generated. Although the device was not helpful, it might be useful to reconsider the use of a similar device, given the skill with which Kanzi and other apes at the LRC manipulate joysticks for other purposes.

Another type of prosthesis is already in use at the LRC—that is, the lexigram boards speak the associated English word when a lexigram is pressed. Although the experimenters often watch, rather than listen to, which lexigram is pressed because the easily portable boards do not talk, the apes are communicating redundantly when using the talking boards, through both visual and auditory media. Other types of vocal prostheses could be developed in the future, for example, a computer interface and program that translated the vocalizations that apes can produce into sounds more similar to English. However, it remains a possibility that special training techniques can be invented that make it possible for apes, most likely bonobos, to use their natural equipment to generate some approximation to human vocal language.

Kanzi is now over 20 years old, a powerful bonobo adult who is much too large and strong to go to the forest. It is clear that he misses all that it offered. He still asks to go to various sites, but now he must ask some caregiver to go for him. He delights in learning through word-lexigrams and hearing what the caretaker says that they will bring back for his eating pleasure.

Panbanisha and Panzee: The incredible success that Kanzi demonstrated, especially in his ability to comprehend, encouraged the researchers at the LRC to immerse bonobos in a language-rich environment when they were younger than Kanzi had been at his first intensive exposure to language. They speculated that the reason for his success relative to Matata was that Matata had been too old and too habituated to the ways of the forest to pay much attention to the effete lexigrams. The researchers also wondered why Kanzi had done so much better than the chimpanzees—could it be that bonobos were genetically more capable of learning language, or was it the way in which he had been raised in a more language-rich environment? Would infant chimpanzees learn in the same way? The researchers decided to raise Panbanisha, a bonobo, with an infant chimpanzee, Panpanzee—Panzee for short.

Raising two infant apes together was a Herculean task. Researchers worked with them for hours every day; there was no such thing as a vacation from them, and they were far more active than a pair of human infants. They slept together in a single bed every night; they had to be inseparable for purposes of comparison. Caretakers commented assiduously and simply on every aspect of their care, coordinating their comments with the use of lexigrams whenever possible.

By the time they were between 2 and $2\frac{1}{2}$ years old, Panbanisha could clearly understand speech. The researchers almost gave up on Panzee at that point, assuming that it was, indeed, a genetic superiority of bonobos for acquiring language that accounted for Kanzi's success. However, they did not give up, and a few months later Panzee showed that she also had a precise understanding of some spoken English words.

Thus both of these young apes learned to comprehend human speech. For both of them, as for human children, comprehension preceded production. They understood, but could not productively use, word-lexigrams. They confirmed the hypothesis that it was the plasticity of a young ape's brain in a language-structured environment that enabled them to acquire important language skills without formal training or differential reinforcement of responses in daily training sessions. Although Panzee was a bit less adept than Panbanisha, that did not invalidate the conclusion that chimpanzees, like bonobos, could acquire language spontaneously through living and observing. Panzee now has demonstrated comprehension of 150 spoken words under controlled conditions. For each word she can select either the thing that it references or the word-lexigram that stands for it. Her understanding of speech is more precise than her understanding of lexigrams, and, unlike Koko, her understanding is handicapped rather than enhanced when two modalities are combined.

Enculturated animals like Kanzi, Panbanisha, and Panzee acquire other skills in addition to those directly involved in the use of lexigrams and the decoding of human spoken language. Tomasello, Savage-Rumbaugh, and Kruger[9] compared the imitative abilities of these three enculturated animals to those of three animals described as mother-reared (Matata, Mercury, and Tamuli) and eight 18-month-old and eight 30-month-old children. Panzee and Mercury are chimpanzees, while the other two animals in each group are bonobos. The authors instructed and trained their subjects to imitate actions and results using external objects, either immediately or after a 48-hour delay. Sixteen novel objects were used in the testing, which involved some simple and some complex imitation.

The results were very clear: the mother-reared chimpanzees and bonobos performed less well than the other three groups at immediate imitation, whether simple or complex. After 48 hours the enculturated chimpanzees or bonobos imitated significantly better than the other three groups (two of them human children), which did not differ from one another. Enculturation clearly enhances significantly the ability of chimpanzees and bonobos to imitate.

The groups were not matched in age, with the enculturated animals older on average than the other subjects, which could account for this group's superiority at delayed imitation. Another variable that was not equated was that the children were tested by unfamiliar adults, and the bonobos and chimpanzees by familiar caretakers, but the enculturated and mother-reared animals did not differ in that respect.

These photographs document the early meeting between Panbanisha and P-Suke described in the text below. They have proved to be very successful parents, and their offspring show great promise as language learners and as integrators of bonobo and human culture.

The authors distinguish carefully between true imitation and other forms of social learning that may account for some apparent cases of imitation and cultural transmission. They conclude that enculturation is critical if apes are to learn to imitate, and that human children probably depend on a similar set of experiences. The enculturated animals probably enjoyed some unavoidable advantages over the mother-reared animals, related to experimental methodology; that is, they were probably more comfortable with humans and human artifacts, and they may have understood the experimental instructions better. However, the differences in their behaviors were so dramatic that it seems unlikely that these advantages for the enculturated group would account for them. More likely, enculturation produced attentional and cognitive changes that made the enculturated animals more able at imitation.

Panbanisha exhibited very surprising behavior on one occasion when I was visiting the LRC. P-Suke, another bonobo, had recently been obtained from Japan as a possible breeding male who would introduce genetic variation into the LRC's group of bonobos. It was uncertain whether he would be a good breeder because he had had little or no prior exposure to other bonobos. He had recently completed his quarantine period and was being gradually introduced to Panbanisha as a possible mate. The two were not yet allowed to be in the same cage.

When Panbanisha was brought into the building where P-Suke was caged, he immediately exhibited an erection. Panbanisha was then allowed to approach his cage, whereupon the two immediately tried to copulate through the mesh of his cage. I could not determine their degree of success, but Panbanisha quickly turned to Sue and the lexigram board. She apparently asked for a wrench; at least Sue gave her one. Panbanisha then returned to the corner of the cage with the wrench and appeared to be trying to unbolt the sides of the cage. I luckily was snapping pictures as this occurred, and the attempts at sex through the mesh and at dismantling the cage are captured in the two photographs above. The reader may decide that Panbanisha's behavior was random, or related to a love of wrenches, but I think it more likely that she wanted to get into the cage with P-Suke and understood the relationship between wrenches, bolts, and cages. Such an understanding would clearly be impossible for an ape that was not enculturated.

Peter Marler's[11] perspective of the research with apes and their skills is succinctly captured in his statement: "The Kanzi project and its predecessors are stunning demonstrations of the powerful influence that the social environment can exert on behavioral development, regardless of whether or not the rubicon of true language is ever crossed."

DR. SAVAGE-RUMBAUGH PRESENTS HER VISION OF THE FUTURE

What is critical at this time is to provide apes with an environment in which their capacities can flourish. While the childhood of our apes was one of freedom of choice, and they had no sense of being captive creatures, this world could not be maintained as they grew older. If they and their offspring remain constrained and confined, their small culture will vanish, as have all such cultures. If this happens we will lose the opportunity to learn how it is that they are coming to speak a Creole of our language, and how their toolmaking skills will ratchet their abilities in other domains such as music, counting, categorizing, and writing.

As adults they need free access to come and go between the world of the forest and the world of the laboratory. They need to do so without the restraint of leads. They need to be able to engage the younger members of their community in a life that requires efforts to locate food, to plan their travel, to communicate these plans, to protect themselves from danger, to pass on their skills of tool use and language, just as was done with them when they were small. In such an environment they might expand their writing to a system of leaving messages like that which I began to document among wild bonobos. (Wild bonobos broke off or pushed down vegetation at points where trails diverged, making certain to indicate the direction or trail that they were taking. The only reasonable explanation for this behavior was that they were attempting to indicate information regarding direction of travel for bonobos who were following them.) Certainly their extensive experience with a written system of communication will alter and flourish in new and unusual ways.

An environment that provides freedom of choice and freedom of spirit is mandatory for adults. It is the only one in which the initial research trajectory laid down for them can continue to flourish. This trajectory provided for them a path of desire to acquire skills on their own in order to become cultural beings with adult competencies. It has been this desire, and the fact that it emerged within a sociocultural context, that sets Kanzi and Panbanisha apart from apes trained as " demonstrators of the intellectual prowess of individual ape minds." For Kanzi and Panbanisha and the other bonobos here, intellect, competency, consciousness, culture, and capacity are a product of a group interacting within a complex environment. This is the natural way of primates.[12,13,14,15]

Kanzi and Panbanisha differ in that they are the only living group of apes who possess large components of human culture in their language, their vocal system, their social lives, and their ways of partitioning the world. Consequently we have much to learn by giving them a future that permits them the dignity they require to carry on lives as competent adults in a social group. They must move away from a world of cages where all their food is provided and where a human caretaker outside the cage must be a facilitator for every decision they make.

If they are to expand their latent potentials, which they are ready to do as adults, their world must be a real one in which they have jobs to do and roles to fulfill in the culture of their community. It is a great loss to knowledge that all of the apes reared in language environments have been confined to small cages upon reaching adulthood because no proper field setting was obtained for them. As adults, they have far more to teach us than we learned from them as children. But they cannot teach us in environments that do not foster choice, initiative, and responsibility, and thereby promote the emergence of latent competencies that were observed as children, but can now be executed as adults. We have seen sparklings of such talents in how Panbanisha and Kanzi respond to Nathan and Nyota. They show them photos and point to specific objects within the photos to draw their attention to them as they vocalize. They look at and point to specific plants and vocalize. They demonstrate tool use as the youngsters watch. But these bits of culture need integration into a larger context in which the important aspects of daily life are guided by their decisions, not those of human caretakers. They need to take the youngsters daily into the forest and show them all that they have learned about travel, locating food, surviving, planning, coordinating, and preparing for the future.

Such an environment needs to provide:

1. A large forested area that will contain both natural foods and foods that can be placed at certain locations. This will permit them to plan and co-ordinate group travel and will provide opportunities upon return for discussion of things that happened in the forest.
2. An adjacent housing area to which the bonobos can return for safety.
3. The presence of special social settings that enhance their opportunities for writing, tool making, music, myth making, and narrative.
4. The capacity for human beings to enter safely into their outdoor environment through protected walkways so that we may continue to learn about bonobos, to observe the processes of cultural transmission and to communicate with bonobos when they wish to communicate. Keyboards would be located within the walkways, to permit the bonobos to invite the visitors to travel to different locations, to share in games of chase, and to encourage interactions around tool making, games of pretend, and storytelling.
5. The natural environment should be coupled with an indoor computer-facilitated world that will permit them to gain access to objects upon request, to open and close off various rooms, to have rooms automatically cleaned, to connect to other apes over an ape-based Internet search engine, to fill pools of water, as well as to communicate with visitors.
6. An opportunity for Kanzi and Panbanisha to expand the vocal communicative competencies that are unique to them in that they can provide vocal answers to questions posed by human beings, to provide, for the first time, an entryway into the vocal language of another species.

REFERENCES

1. T. Kano, *The last ape: Pygmy chimpanzee behavior and ecology* (Stanford University Press, Stanford, CA, 1992).

2. H. Davis and D. Balfour (Eds.), *The inevitable bond* (Cambridge University Press, Cambridge, 1992).

3. E. S. Savage-Rumbaugh, *Ape language: From conditioned response to symbol* (Columbia University Press, NY, 1986).

4. E. Bates, *Language and context* (Academic Press, NY, 1976).

5. D. Premack and G. Woodruff, Does the chimpanzee have a theory of mind? *Behavioral and Brain Sciences*, **4**, 515–526 (1978).

6. M. A. Romski and E. S. Savage-Rumbaugh, Implications for language intervention research: A nonhuman primate model, in: *Ape language: From conditioned response to symbol*, edited by E. S. Savage-Rumbaugh (Columbia University Press, NY, 1986).

7. F. de Waal and F. Lanting, *Bonobo: The forgotten ape* (University of California Press, Berkeley, 1997).

8. E. S. Savage-Rumbaugh and Roger Lewin, *Kanzi* (Wiley, NY, 1994).

9. M. Tomasello, S. Savage-Rumbaugh, and A. C. Kruger, Imitative learning of actions on objects by children, chimpanzees, and enculturated chimpanzees. *Child Development*, **64**, 1688–1705 (1993).

10. J. P. Taglialatela, S. Savage-Rumbaugh, and L. A. Baker, Vocal production by a language-competent bonobo, *Pan paniscus. International Journal of Primatology*, (in press).

11. Peter Marler, How much does a human environment humanize a chimp? *American Anthropologist*, **101**(2), 432–436 (1999).

12. S. Savage-Rumbaugh, W. Fields, and J. Taglialatela, Language, speech, tools and writing: A cultural imperative, in: *Between Ourselves: Second-person issues in the study of consciousness*, edited by E. Thompson (Exeter, UK: Imprint) 273–292 (2001).

13. E. S. Savage-Rumbaugh, and W. Fields, Linguistic, cultural and cognitive capacities of bonobos (*Pan paniscus*), *Culture & Psychology*, **6**(2), 131–153 (2000).

14. E. S. Savage-Rumbaugh, W. Fields and J. Taglialatela, Ape consciousness–human consciousness: A perspective informed by language and culture, *American Zoologist*, **40**(6), 910–921 (2000).

15. E. S. Savage-Rumbaugh, and W. Fields, Language and culture: A transcultural interweaving, *Language Origins Society*, **28**, 4–14 (1998).

CHAPTER ELEVEN

Chantek the Beautiful

A visit to Lyn Miles's laboratory these days is a visit to Atlanta Zoo. It was not always that way; Lyn worked with Chantek in Chattanooga for 8 years, when things were going smoothly for her at the University of Tennessee.

Dr. Lyn Miles got her Ph.D. degree in anthropology from the University of Connecticut in 1974, although she took many of her classes at Yale; the two universities had an exchange arrangement. She did her dissertation research at William Lemmon's Institute for Primate Studies at the University of Oklahoma. Her interest in the social origins of meaning led her to an intense interest in animal language research. She arranged a 2-year visit to Lemmon's facility during 1973–74. There she did dissertation research, primarily with the chimpanzees Ally, Booee, and Lucy.

Dr. Miles found that 25 percent of chimpanzee sign exchanges were about topics that did not involve food and drink. Her dissertation thus disproved the allegations of some critics that all chimpanzee signs were attempts to obtain food or drink, while verifying that a large proportion of them were so motivated.

Soon after receiving her degree, Dr. Miles received an invitation to a job interview at Georgia State University, where Dr. Duane Rumbaugh had recently become Chair of the Department of Psychology. However, on the very day she arrived all hiring at the university was frozen. Thus she left empty-handed, except for Dr. Rumbaugh's advice concerning animal language, "Try orangutans"!

Lyn was hired by the University of Tennessee at Chattanooga soon after her futile visit to Georgia, and the University supported her attempt to train an orangutan in pidgin American Sign Language. The Yerkes Laboratory of Primate Biology allowed her to choose an orangutan from its population. Lyn started to observe Chantek (which means "beautiful" in Malaysian) when he was 6 months old. He was born on December 17, 1977, at Yerkes. His

Bornean orangutan mother was Datu, and his Sumatran orangutan father was Kampong. When he was 9 months old, Lyn took Chantek to Chattanooga to begin cross-fostering him. He lived in a five-room trailer inside an enclosure next to the music conservatory on the campus. Dr. Miles and several student co-workers spent 24 hours a day with him at first, acclimating him to the trailer with toys and climbing ropes, and the outside enclosure with ropes strung between trees. Orangutans are very arboreal animals, and Chantek's environment was designed to accommodate the orangutan way of life, as well as the human. After about a year, in November of 1979 when he was nearly 2 years old, Chantek started his sign language training.

After the first 8 years with Dr. Miles, Chantek fell upon evil times. He escaped from his enclosure, even though it was topped with electrified wire. He went into the administration building, where he was found "making a phone call" (Chantek liked to listen on the phone, and had apparently picked

In addition to his substantial language skills, Chantek became a spontaneous macame' and necklace maker. He ties knots where strategically necessary to keep the beads from slipping off. He threads the beads by deft coordination of his hands and ties knots with his lips and tongue in well-coordinated movements. When he decides that his product is "done," he hands it back to Lyn Miles. It is, indeed, a remarkable and accomplished skill that he acquired without training. Photo by courtesy of Dr. Lyn Miles.

up a phone to listen). The Chancellor at Chattanooga was less than charmed and, instead of congratulating Chantek on his intelligence and escape artistry, seized the opportunity to get rid of him. According to Dr. Miles, the Chancellor was "afraid of dogs, cats, and goldfish," and had no appreciation for research on animal language. So Chantek was shipped back to Yerkes where, for a time, Dr. Miles was allowed to continue to work with him. But soon the political winds shifted, and she was told that Yerkes was going to "put the animal back in Chantek." Whether or not they did that remains uncertain, but it is certain that they put an additional 250 pounds *on* Chantek, who ballooned to 500 pounds while restricted to a small cage and spoiled with too much food and too little exercise. It is not clear how anyone could believe that such a life would "put the animal back in him," but, in any event, for several years Dr. Miles, who had for years spent approximately 40 hours per week with Chantek, was not allowed even to see him.

Finally, a reporter from CNN (the cable news network) threatened to reveal the conditions under which Chantek was living, including the separation from Dr. Miles. Yerkes relented to the extent that they allowed her to see Chantek from a distance of 20 feet while wearing a mask. Sign language training is impossible under those conditions. However, Chantek was later transferred from the Yerkes laboratory to the Atlanta Zoo. The zoo director, Dr. Terry Maple, has transformed the Atlanta Zoo from a third-rate to a first-rate facility, and is working to develop a computer simulation of Chantek's signing.

At the zoo Lyn was allowed to help acclimate Chantek to the presence of a female orangutan, who is now his cage mate. Lyn has started to work with Chantek again, and to reintroduce him to the use of sign language. She no longer has to wear a mask. It has been difficult because Chantek had to be put on a diet to get him back to a normal male orangutan weight of about 240 pounds. Thus he is always hungry, and only wants to talk about food. In addition, Lyn is not allowed to touch Chantek. This poses a problem because teaching an orangutan is usually a very physical process, sometimes involving a wrestling match and arm locks, along with hugs and squeezes!

A VISIT TO CHANTEK AT THE ATLANTA ZOO

Dr. Miles arranged for Duane Rumbaugh, my son, Gaines Hillix, and me to visit Chantek at the Atlanta Zoo on Saturday, December 13, 1999. We were very pleased, because the zoo people are hesitant to let people visit Chantek.

Dr. Miles showed us videotapes of Chantek as a young orangutan in Chattanooga. All were interesting, but the most impressive was a tape of a young Chantek the first time he was handed a piece of flint and a large elliptical hammer stone. With no training whatever, he immediately started pounding on the flint with the hammer stone. That was amazing enough, but even more amazing was his persistence. He kept pounding away for at least 5 minutes; when he produced a tiny flake, he brushed it aside, but when he produced a larger one, he kept it. It was as though he wished instinctively to produce a cutting edge. He finally gave up; the flint was a very hard, small piece. The surface was far from optimal for pounding, and his pounding was less effective than one might expect from such a strong animal. It will be fascinating to observe what he does with more appropriate materials and some demonstrations, or other forms of training. He seems to have as much native talent as Kanzi for flint-knapping. That is not surprising; an orangutan named Abang learned to make stone flakes via imitation,[1] and then used a flake to cut a string that held a box closed, the precise problem that Kanzi later solved under Toth's tutelage.

After the videos and discussions, we went around the corner to Chantek's cage. Chantek is pronounced like a combination of the Chan in Charlie Chan and the tech in high tech. He no doubt looks strikingly handsome to female orangutans, as his Malaysian name indicates he should be. He has the prominent cheek pads of the adult orangutan, small eyes that appear to be deep in his head, and a rich coat of orangutan-red hair that looks something like carroty dreadlocks.

Lyn warned us that Chantek might be somewhat upset by the appearance of three strange males. Orangutan males in the wild space themselves widely and defend their territories. Lyn had tried to tell Chantek that three new friends were coming, but there is no way to know for sure that he got the message. We got a brief introduction at the near side of the cage, and then walked around to the far side, where Lyn had gathered her materials—semiprecious jade with small holes drilled in it, string, apples, yogurt, and some sticks about $\frac{1}{4}$ inch in diameter and 18 inches long. On the way around the cage, Chantek got as close as he could to us and spit on Gaines. Lyn said that, to her knowledge, Chantek had never spit on anyone before. However, Gaines weighed about 275 pounds, and had a beard that probably made him appear more threatening than most males. Chantek clearly did not like him!

When we reached the far side of the cage, Lyn demonstrated how Chantek would play the game Simon Says. She instructs him to do so by signing with the forefingers of the two hands together and pointing at Chantek, with the other fingers and thumb fisted. Then she could hold up a leg, or whatever, and Chantek would follow suit. Then Lyn made the sign again, but pointed at Gaines, indicating that Chantek should imitate Gaines. Chantek mulled this instruction for perhaps 10 seconds, and then turned around and walked away! Lyn had to issue a lengthy apology to get Chantek to come back. Later she got Chantek to imitate Duane and me, but she didn't try Gaines again.

Lyn then gave Chantek a piece of the decorative jade with a hole drilled in it, together with a sturdy piece of string. Chantek put the jade and string in his mouth, moved his lips, tongue and fingers about for a bit, and then took out the jade, miraculously strung on the string, and with a knot in it! He proceeded with his huge fingers to tighten the knot, and hand the result back to Lyn. The process was repeated with two more pieces, after which Lyn handed Chantek the stick and one of the strung pieces. Chantek tied it onto the stick, and then tied on the second and third piece. The result looked a bit like a small wind chime and made a fine decorative piece. I have been told that some humans can duplicate Chantek's knot-tying feat with their tongues, but I doubt that their ability is as great as Chantek's. Thus I doubt that apes' difficulty with auditory language is related to insufficient control of the tongue!

Chantek was rewarded periodically with an apple. No doubt, as stated above, he is chronically hungry as a result of being reduced to and maintained at a proper weight. All human dieters can sympathize with him. Chantek also indicated that he wanted a yogurt by pointing to it, and then a second yogurt.

I am surprised that so many animals are willing to do the bidding of humans; we are smaller, and weaker by far, than many animals (think elephant), and they would seem to have little reason to want to please us. Perhaps they do so primarily because they are rewarded with food and drink, but I do not think so. Dogs, for example, love to be petted, and primates are apparently fond of being groomed. Dr. Louis Herman even uses social reinforcement with dolphins, and Alex, Dr. Irene Pepperberg's African Gray parrot obviously enjoys human companionship. A remarkable number of animals seem to enjoy human approval.

At the end of our interaction, we walked by the cage in the reverse direction; Chantek again followed over to the side and spit on Gaines yet again. Neither spit job was more than a very thin explosion of saliva, so we all, fortunately including Gaines, thought it was pretty funny. Lyn was embarrassed, but Chantek was not.

CHANTEK AS A PERSON

Dr. Miles makes a persuasive case that Chantek fulfills the criteria for person-hood, and thus deserves rights and protection similar to that accorded to other persons. Chantek has several properties that we associate with being a person. Among them is at least a limited intellectual capacity; Chantek's intellect is comparable to that of a human 2- to 3-year-old,[2] to whom we accord the rights of personhood (some estimates have even put his thinking at the level of a 4-year-old human). A book in which a chapter by Dr. Miles appears, *The great ape project*,[3] was written in an effort to justify rights for all the non-human great apes. The book begins with the following declaration:

We demand the extension of the community of equals to include all great apes: human beings, chimpanzees, gorillas, and orangutan.

"The community of equals" is the moral community within which we accept certain basic moral principles or rights as governing our relations with each other and enforceable at law. Among these principles or rights are the following…(p. 4).

The editors and contributors go on to list three basic rights, the right to life, the protection of individual liberty, and the prohibition of torture. These rights are not always achievable; for example, it may not be possible to give animals born and raised in zoos or other captive situations their individual liberty; the variation on Patrick Henry would be "If you give me liberty, you give me death." But the declaration would forbid taking any great apes captive in the future. In interpreting the declaration, we should note that the authors intended bonobos to be included as a type of chimpanzee, although we have noted that they are a separate species rather than a subspecies.

The positions expressed in *The great ape project* are similar to those expressed later in Steven Wise's (2000) *Rattling the cage*. Wise, a lawyer, puts more emphasis on legal rights, but he opposes experimentation and supports legal protection of all animals who demonstrate intelligence, at least limited self-awareness, and the ability to act intentionally. He believes that all the great apes, dolphins, elephants, and African gray parrots clearly deserve legal protection. However, in opposing experimentation he seems to mean experimentation that harms the animal, not, for example, animal language experiments on animals that are already captives.

In addition to his intelligence, Chantek has consistently manifested self-awareness since he was $3\frac{1}{2}$ years old. The best-accepted formal test of self-awareness is Gallup's "mark test." A mark is placed on the forehead of an anaesthetized, or otherwise unaware, animal. The awake and alert animal is then shown a mirror. If he or she immediately touches the mark, the test is passed. The testee has shown by his or her response that he or she is aware that the image in the mirror is herself or himself. Animals that do not have self-awareness respond to the image as though it were another individual; some animals display at the image; others reach behind the mirror, which is a more insightful response but not a self-aware response. Dr. Miles tested Chantek in this way repeatedly, watching his development of self-awareness until it became consistent at $3\frac{1}{2}$. After that time he actively sought out mirrors in order to groom himself, and consistently named his image in the mirror CHANTEK.[4] When he was 7 years old he even tried to curl his eyelashes, using the mirror, as he had seen his caretaker do. Thus, contrary to earlier claims, bonobos, chimpanzees, gorillas, and orangutans that have received language training all present clear evidence of self-awareness.

Chantek also appears to imitate and to deceive his keepers (an all too human characteristic), and he has learned to use sign language as a means of communication. Lyn's argument for Chantek's personhood has a strong foundation!

CHANTEK AND AMERICAN SIGN LANGUAGE

During his years at Tennessee Chantek had three or four primary keepers, and a total, including student volunteers, of about a dozen caretakers. His language training was naturalistic. Dr. Miles avoided the systematic and repetitive procedures used for Terrace's chimpanzee, Nim. Chantek was enculturated through normal interactions with his human companions, rather than trained. In this respect Chantek's education was rather similar to Washoe's in the later stages of her training by the Gardners, once they had rejected Skinnerian operant techniques as their primary training method. Chantek, like Washoe, learned sign primarily through imitation and molding (having his hand or hands placed in the correct position and moved appropriately). Chantek surprised his caretakers by offering his hands for molding as a way of asking for the name of unknown objects. In another exhibition of creativity, Chantek invented five new signs of his own, including a very iconic holding his hands to his lips while blowing for balloon.

In all, Chantek learned a total of about 150 signs over an 8-year period, 127 of which met the Gardners' criterion for acquisition. Chantek had to use each new sign correctly on at least half the days of a month in order for it to be counted as an item in his vocabulary. That is equivalent to the "P criterion" that Dr. Francine Patterson used with Koko. Chantek's vocabulary was comparable to Washoe's (Washoe had learned 132 or 133 signs at the age of 51 months, and 170 signs after she was transferred to the Foutses' care). However, the Gardners' criterion was more demanding; Washoe had to use each sign on 14 *consecutive* days if it were to be counted in her vocabulary. On the other hand, Terrace (1979) required Nim to use a sign on only 5 consecutive days. All the criteria are arbitrary, of course, but their inconsistency makes comparisons between individuals and species confusing and difficult. This difficulty is compounded by differences in training methods and in the age at which each animal starts the learning process.

Dr. Miles describes Chantek's language training as follows:[5]

Chantek's sign language training was broken down into three stages. First, we established the rules for human communication and discourse; second, we taught him specific gestural signs which we brought to criterion for acquisition; and third, we sought to encourage his independent use of signs to transmit meaningful information about himself and his environment (p. 48).

Under this regimen, Chantek's progress was quite steady. A graph of his vocabulary size as a function of time looks nearly linear over the period of his stay in Tennessee. The graph[6] ends in 1986, when Chantek's return to Yerkes essentially ended his language training. After his first 2 years of training, Chantek both saw sign and heard English. As a result, he later was able to produce the correct signs for English words; in this respect, his training and performance paralleled Koko's. Dr. Miles says that Chantek now understands hundreds if not thousands of English words. As in the case of Koko, there has been no precise determination of how many words Chantek comprehends, and such a determination may not be practicable.

Chantek's signs were much more deliberate than those of the chimpanzee, Ally, with whom they were compared (0.92 seconds per sign versus 0.39 seconds per sign). It is a good bet that the average gorilla sign duration would lie between the duration for the frenetic chimpanzee and the duration for the deliberate orangutan. After he had learned the meaning of *sign better*, Chantek would respond to that request by trying to make an even more careful and deliberate sign.

Miles[5] compared Chantek's use of sign language most extensively to that of Nim Chimsky, no doubt because Terrace had been so skeptical about ape language after his close study of Nim Chimsky's use of signs. Terrace found that a high percentage (about 40 percent) of Nim's signs were imitations of immediately preceding signs made by his trainers. Only 3 percent of Chantek's signs were imitations. At the other end of the continuum of independence of signing, only 8 percent of Nim's utterances, but 37 percent of Chantek's, were "spontaneous." Spontaneous utterances are defined as those that are not immediately preceded by a caretaker's utterance. Intermediate types of utterance are expansions of a prior caretaker's utterance (6 percent for Chantek, 8 percent for Nim) and novel utterances (those that repeat none of the caretaker's utterance; about 45 percent of Nim's utterances and about 35 percent of Chantek's fell into this category).

The differences between Nim and Chantek are striking, and are more likely to be consequences of the different techniques used to train them than of their different species. However, given that less than half of Nim's utterances were imitations, it is difficult to understand the widespread perception that Nim only imitated. According to Roger Fouts,[7] Nim's signs were much more spontaneous and less imitative after his move to Oklahoma than they had been in the situation where he was trained and tested. Chantek almost never imitated; this low rate of imitation may indicate that Chantek's early training in the rules of conversation helped him to understand that each party to a conversation was to add new information.

Another feature of Chantek's signing provides additional proof that he is not imitating: Chantek signs even when he is alone with people who do not know sign. Although he does not indicate it, he is probably frustrated at the stupidity of his non-signing companions.

Miles' analysis of the properties of Chantek's utterances led her to believe that his signing behaviors were at the level of a Stage 1 human child. Stage 1 children imitate at a higher rate (18 percent) than Chantek, and their spontaneous utterances are at a comparable level (about 31 percent on average).

The mean length of Chantek's utterances (MLU) never became very impressive, topping out under 2.0, but he did make reasonably continuous progress over the first 17 months of training, beginning at 1.00 and ending at 1.91. This MLU, like his spontaneity, is sufficient to put him at Stage 1 of human language development. It is difficult to compare Chantek's MLU with Nim's; Terrace reported that Nim's MLU stopped increasing at about 1.6, although Nim emitted long strings (up to 16) of repetitious, mostly nonsensical signs.

CHANTEK AND THE USE OF SIGNS FOR DECEPTION

Miles[8] believes that Chantek used signs for deception, and that such use indicates more sophisticated cognition than the use of signs for straightforward requests and comments. Between January and July 1982, she recorded all instances of Chantek's behavior that might be deceptive. She recognizes the difficulty of distinguishing deception from simpler behaviors like simple errors or overgeneralized use of signs. Eighty-seven possible instances were recorded. Chantek often used the sign for dirty in ways that might have been deceptive; that is, the sign was supposed to mean that Chantek needed to eliminate, but on about 40 occasions he failed to do so. A skeptical interpretation was that Chantek mistakenly thought he had an urge to defecate, or simply made the wrong sign. However, the bathroom was off limits except when it was to be used for its intended purpose, and it contained fascinating objects for play, like toilet paper, faucets, and dials on a washing machine and dryer. Because Chantek often tried to take advantage of the play opportunities afforded by the bathroom, it appeared to less skeptical observers that Chantek used the sign dirty to deceive.

Chantek exhibited several other types of behavior that also appeared to be deceptive. He often signed CAT when his caretakers saw no animal, apparently to distract attention from testing or to avoid returning to the trailer when on an outing. Although, as Dr. Miles points out, it is nearly impossible to rule out alternative explanations of behaviors that appear deceptive, the large number of these behaviors, considered in context, makes it very likely that at least some of them were intentionally deceptive. In a few cases it appears that Chantek both intended to deceive and realized that his caretaker might uncover his deception. For example, he bit an eraser off a pencil and then signed FOOD, EAT to his caretaker while opening his mouth. He apparently had palmed the eraser, because he was later caught playing with it. Such behavior borders on demonstrating that Chantek has a "theory of mind," although Dr. Miles explicitly denied that apes have such a theory.[8] Similar deceptive behaviors have been reported in chimpanzees and gorillas.

CHANTEK'S ABILITY TO IMITATE

The belief that apes can imitate is so much a part of our culture that "to ape" is synonymous with "to imitate." Given that belief, it is surprising that true imitation is so difficult to demonstrate scientifically. However, Chantek provided so many examples of apparent imitation that it is impossible to escape the conclusion that he was an enthusiastic, and perhaps unusually talented, imitator. One example was his use of the sign for imitation, DO-SAME, which he had never been taught to make. He used the sign to reverse the roles of himself and a caretaker; the caretaker had been asking Chantek to imitate various actions, when Chantek very carefully and correctly signed DO-SAME while maintaining eye contact with the caretaker. That was a kind of meta-imitation, at the same time that it showed an ability to shift perspective.

THE SIGNIFICANCE OF THE CHANTEK PROJECT

The most obvious result of the Chantek project has been that it destroyed the last vestiges of support for any assumption that the slow and deliberate orangutan is also the stupid orangutan, relative to the other great apes. Some people made that assumption because the orangutan line diverged from the line that led to humans earlier than did the gorillas (who diverged next) and the chimpanzees and bonobos (who diverged last). The egocentric human view was that the closer you were to humans genetically, the smarter you must be!

It is sobering to recognize that, the best we can now tell, that isn't true. There is insufficient evidence to construct a great chain of intelligence among bonobos, chimpanzees, gorillas, and orangutans. The intelligences of these species apparently increased, or perhaps decreased, at about the same rate after they separated from the hominid branch as they had while joined to the branch. Previous objective tests of ape intelligence by Rumbaugh and McCormack[9] had already indicated that the great ape groups (Gorilla, Pan, and Pongo) did not differ in their ability to form learning sets, and studies of the ability to transfer old learning to related new situations also failed to reveal significant differences between the groups of great apes.

Chantek developed linguistic skills in a manner strikingly parallel to the way deaf human children and other great apes learn sign language. In the nonhuman great apes the number of words in the vocabulary and the mean length of utterance increase more slowly and to a lower maximum than is the case with human children, but the sequence of events is similar. Thus Chantek added to the previous evidence indicating that the nonhuman apes share with humans significant cognitive underpinnings for the comprehension and production of language.

Finally, Chantek's abilities to imitate actions (including signs), to be deceptive, to be creative, and to be self-aware, make him, like Koko, Kanzi, and many other language-trained apes, appear to be almost human.

Dr. Miles is continuing her work with Chantek. In addition R.W. Shumaker and Karyl Swartz[10] are studying the cognitive capabilities of two Orangutans at the Think Tank of the National zoo in Washington, D.C., an agency of the Smithsonian Institution. Drs. Shumaker and Swartz are using a symbol–embosed keyboard, reminiscent of the Lana keyboard with arbitrary symbols as words for language studies and Arabic numerals for studies of counting. The investigators are conducting innovative studies of memory, as well as of cognition.

REFERENCES

1. R. V. S. Wright, Imitative learning of a flaked tool technology—the case of an orangutan. *Mankind*, 8(4), pp. 296–306 (1972).
2. H. L. Miles, Language and the orangutan: The old "person" of the forest, in: *The great ape project*, edited by P. Cavalieri and P. Singer (St. Martin's Press, NY, 1993), pp. 42–57.
3. P. Cavalieri and P. Singer, *The great ape project* (New York: St. Martin's Press, 1993).
4. H. Lyn Miles, Chantek: The language ability of an enculturated orangutan (*Pongo pygmaeus*), in: *Proceedings of the international conference on "Orangutans: The neglected ape"*, edited by J. Ogden, L. Perkins, and L. Sheeran, (Zoological Society of San Diego, San Diego, 1994), pp. 209–219.
5. H. Lyn Miles, Apes and language: The search for communicative competence, in: *Language in primates: Perspectives and implications*, edited by J. de Luce and H. T. Wilder (Springer-Verlag, NY, 1983), pp. 43–61.
6. H. L. Miles, The cognitive foundations for reference in a signing orangutantan, in: *"Language" and intelligence in monkeys and apes*, edited by S. T. Parker and K. R. Gibson (Cambridge University Press, Cambridge, 1990), pp. 511–539.
7. R. Fouts, with S. T. Mills, *Next of kin* (William Morrow and Co, NY, 1997).
8. H. L. Miles, How can I tell a lie? Apes, language, and the problem of deception, in: *Deception: Perspectives on human and nonhuman deceit*, edited by R. W. Mitchell and N. S. Thompson (State University of New York Press, Albany, NY, 1986), pp. 245–266.
9. D. Rumbaugh & C. McCormack, The learning skills of primates: A comparative study of apes and monkeys, in: *Progress in primatology*, edited by D. Starck, R. Schneider, and H.-J. Kuhn (Gustav Fischer Verlag, Stuttgart, 1966), pp. 289–306.
10. Shumaker, R. W. & Swartz, K. 2002. When traditional methodologies fail: Cognitive studies of great apes in *The Cognitive Animal* (Bekoff, M.; Allen, C.; Burghardt, G. eds.), MIT Press, pp. 335–343.

CHAPTER TWELVE

Ai Project: A Retrospective of 25 Years Research on Chimpanzee Intelligence

by Tetsuro Matsuzawa

AN APE-LANGUAGE STUDY BEGINS IN JAPAN

In November 1977, a one-year-old female chimpanzee arrived at the Primate Research Institute of Kyoto University (KUPRI), Japan. She was wild-born, probably in Sierra Leone, West Africa, and was purchased through an animal dealer. Importing wild-born chimpanzees was still legal at the time, as Japan only ratified the Convention of International Trade in Endangered Species (CITES) four years later, in 1980. In the 1970s, Japan imported more than 100 wild-born chimpanzees, mainly for biomedical research of Hepatitis B. This infant chimpanzee was one of them. However, instead of being sent to the biomedical facilities, she was sent to KUPRI, where she was to become the first subject of an ape-language research project in the country.

The chimpanzee was soon nicknamed Ai (pronounced *eye*). Ai means "love" in Japanese, and is also one of the most popular girls' names in Japan. She was estimated to have been born in 1976, and was about a year old at the time. After being examined in quarantine, she was kept in a basement room, only about 4×4 m in size and without any windows. I was 27 years old at the time, the youngest assistant professor in the institute, and was expected to become Ai's principal trainer. I first met her in that dimly lit basement room, a bulb hanging from the ceiling. When I looked into this chimpanzee's eyes, she looked back into mine. This amazed me—the monkeys I had known and had worked with never looked into my eyes. For them, staring straight into one's eyes carries a threat, and they would probably respond by opening their mouths and threatening you back or by presenting their backs and assuming a submissive posture. I had simply thought that chimpanzees would be big black monkeys. She, however, was no monkey. She was something mysterious.

Soon after Ai's arrival, she was joined by two other infant chimpanzees of about the same age. One was a $1\frac{1}{2}$ year old male, nicknamed "Akira," and the other a $1\frac{1}{2}$ year old female, nicknamed "Mari." The construction of the chimpanzee facility had also been completed by this time, and it consisted of four individual residential rooms $(1.5 \times 1.5 \times 2\,\text{m}$ high) and an attached outdoor pen (about 15 square m). The three infant chimpanzees moved to the new facility.

HISTORICAL BACKGROUND AND RATIONALES OF THE AI PROJECT

What is uniquely human? This question has long attracted psychologists. Specifically, many have tried to explore the human mind through comparisons with the mind of the chimpanzee.[1] A big turning point arrived in the early 1970s in this research area: the beginning of the so-called "ape-language" study.

The chimpanzee project in KUPRI started in 1977, originally with the aim of becoming an ape-language study. By the second half of the 1970s, before our project began, three successful and different approaches to ape-language study had been devised: American Sign Language (ASL),[2] Plastic Sign language,[3] and a computer-controlled lexigram system.[5,6] All three projects produced reports that appeared in the journal *Science* and had already received wide attention.

The chimpanzee project in KUPRI was led by Dr. Kiyoko Murofushi, an associate professor and the head of the psychology section at the time. Dr. Murofushi was flanked by three assistant professors, Toshio Asano, Shozo Kojima, and me. However, when the project started, Asano was in the middle of a 2-year sabbatical at the University of California, San Diego (UCSD) and Kojima was about to leave for his 2-year sabbatical to the National Institutes of Health (NIH). Therefore, I had to face the three chimpanzees by myself, as the principal trainer/researcher under the supervision of Dr. Murofushi. Having got my job in 1976, I had only very limited experience in experimental psychology: human visual perception, physiological-psychological study of rats' memory, and only a single year's experience with visual discrimination learning in Japanese monkeys.

I had an intrinsic interest in and motivation to study the visual world of nonhuman primates. How do these animals see the world? Is their perception similar to humans'? Such questions were originally proposed by von Uexkuell, a 19th century ethologist, and were first tackled by a new discipline that emerged in the 1960s called "animal psychophysics." Instead of concentrating on communication and language in the framework of ape language studies, I planned to focus on the study of the perceptual world of chimpanzees by applying the discrimination learning paradigm.

Prior to the project, we had very little experience with chimpanzees. A single female chimpanzee, named Reiko, had been living in the institute since 1968, and had principally served as the subject for a bipedal locomotion study by morphologists. The only psychological research that had been carried out on the chimpanzee explored an operant response for lighting her room. She had a switch with which she could freely turn the lighting on and off, and the researchers recorded the circadian rhythm in the spontaneous switching on and off behavior.

Murofushi and her colleagues, consulting experts in related disciplines, endeavored to clarify the goals and methods of the ape-language research, as well as its stance in relation to the ongoing projects in the USA. The consultants included Dr. Kisou Kubota (neuroscience of frontal lobe of the brain), Dr. Makoto Nagao (computer processing of human natural language), and Drs. Akio Kamio and Susumu Kuno (linguists of generative grammar).

Carefully examining the achievements and the methods of the American studies, we set up our own goals and methods as follows. The goal was to study the acquisition process of an artificial language and the corresponding brain mechanisms. Although it became clear that the apes could learn to use visual symbols to some extent, the acquisition process remained unclear. No one had previously tried to connect the psychological facts to brain functions. There were no Positron Emission Tomography (PET), no Functional Magnetic Resonance Imaging (fMRI), and no magnetic stimulation techniques available. Dr. Murofushi, the leader, was carrying out split-brain research in monkeys in collaboration with Dr. Kubota.[4,8] But already by this time the notion to perform invasive brain studies with the chimpanzees was rejected by everyone involved, and cranial cooling of a hemisphere seemed a suitable candidate as a noninvasive and reversible technique. However, the technique was actually never tried in the real project. We focused instead on the perceptual and cognitive basis of language-like skills mastered by the chimpanzees. In my personal perspective, what I really aspired for was to explore the perceptual world of chimpanzees through clearly defined visual symbols. My central questions were: How do chimpanzees perceive this world? Do they perceive it like we do?

From among the three existing approaches, we chose to adopt the computer-controlled lexigram system developed by Rumbaugh and his colleagues.[5]

Our decision was influenced by three factors. First, we had already established sophisticated computer-controlled experiments involving various visual discrimination learning tasks in monkeys, such that we were able to apply immediately our existing techniques to the new ape-language research project. Second, we had aimed to clarify the acquisition process or underlying perceptual capabilities of language-like skills, which meant that we needed very objective, precise, and detailed records of what we had done and how the chimpanzees behaved. For that purpose, a computer-controlled system was essential for the study. Third, the project had a perspective for future applications of techniques from brain science, and for that purpose we hoped that the subject would sit quietly on a bench facing the computer system.

The computer system that we had used for monkeys in the 1970s was based on DEC PDP12 and PDP8. However, DEC PDP/V03 soon became available for the chimpanzee research. Toshio Asano, in collaboration with technician Sumiharu Nagumo, was in charge of developing the PDP-minicomputer-based system for controlling visual symbols. They originally made an interfacing device for a keyboard with the IEE in-line projectors displaying visual symbols, and also developed a computer program in BASIC programming language. The appearance of the experimental setting was very similar to that invented by Rumbaugh's Language ANAlogue (LANA) project.[6] Based on published articles from the latter, one could easily replicate the setup, even though we had had no contact at all with Rumbaugh's group before our project began. Although our apparatus resembled the apparatus used in their previous study, our goal was quite different and unique. We were attempting to clarify the acquisition process of the symbols and to illuminate clearly the steps involved in complex language-like skills in chimpanzees.

For us, Lana chimpanzee's computer-performance seemed to be a sequence of visual discriminations similar to that in monkeys. For example, when the chimpanzee touches five keys on the keyboard consecutively, the sequence does not necessarily correspond to a sentence such as "Please give Lana chocolate period" in human verbal behavior. We were hoping to investigate how specific visual symbols such as those representing the names of individuals, objects, actions, and so forth, could be established in chimpanzees.

FOCUSING ON PERCEPTUAL AND COGNITIVE PROCESSES RATHER THAN LINGUISTIC SKILLS

It was April 15th, 1978. The chimpanzee Ai participated in a computer task inside an experimental booth ($2 \times 2 \times 2$ m) for the very first time. Her first task was to touch a lit key on the keyboard. Ai was alone inside the testing booth while the experimenter remained outside. There was no direct interaction between her and the human tester. Instead of demonstrations by a model or molding the hands of the chimpanzee, we used the "successive approximation" method for shaping the key-touch behavior. When Ai faced the apparatus, a chime rang and a piece of apple was delivered to a food cup attached to the keyboard. The criterion for delivering the reward was gradually changed, until step by step she began to approach the key, then eventually press it. Since then, Ai has continued to press keys and thus to reveal many aspects of chimpanzee intelligence. Throughout the later stages, the project was characterized by a highly automated computer-controlled system without any social interaction during the tests. However, we made sure that a lot of interaction took place during out-of-test play situations, which produced an affectionate tie between the tester and the subject, and helped to carry out the tests smoothly.

The first paper about the project, in English, appeared in 1982.[8] It described our research system, and the process through which the three chimpanzees acquired visual symbols corresponding to object and color names. The visual symbols, uniquely devised by us, became known as the Kyoto University Lexigram system (KUL). The lexigrams had no background color such as those used in Yerkish Lexigram devised by Rumbaugh and his colleagues. Instead, the lexigrams consisted of black and white patterns and were fundamentally similar to Kanji (Japanese-Chinese characters).

The original keyboard contained three panels. Each panel consisted of 35 keys arranged in 7 rows and 5 columns. The lexigrams were drawn on a film sheet and inserted over the keys. Each key, 1.5×2.0 cm in size, could be back lit, indicating those available in a particular test. Touches to lit keys produced a feedback sound (click), the key light faded, and a facsimile of the lexigram appeared on the inline projectors above the keyboard. There was a display window $(20 \times 30 \text{ cm})$ through which the tester showed an object to the chimpanzee, who sat voluntarily on the bench, facing the apparatus.

The general procedure was matching-to-sample (MTS). In a sense, from the beginning our project was not aimed to become another study of language-like or communication-like interactions with the chimpanzees, in the mold of previous studies. Instead, we attempted to clarify the acquisition processes behind such skills and to illuminate the chimpanzees' capability in their matching skills in a very objective way. Hence, our tasks were exact equivalents of the MTS tasks given to other subjects—monkeys, rats, pigeons—in the laboratory.

After gathering data on color perception, shape perception, and visual acuity, we proceeded to study the recognition of numbers by the chimpanzees. Ai became the first chimpanzee to use Arabic numerals to represent quantities.[7] We still maintain this line of research, focusing on the perceptual and cognitive capabilities of chimpanzees.[8-20] The project increasingly covers short-term memory, the learning of sequences, face recognition, auditory-visual intermodal matching, global-local processing, the perception of moving images, the perception of visual illusions, visual search assymetries, the representation of symbols, categorization, and computer-assisted drawing. Our experimental paradigm continues to offer a unique window into the perceptual and cognitive world of the chimpanzee.

FIELD EXPERIMENTS WITH WILD CHIMPANZEES

Japan is a special country in terms of the study of nonhuman primates. There is an indigenous monkey species in Japan, while other leading industrial nations have none: there are no nonhuman primate species in North America and Europe. This natural precondition has promoted the study of nonhuman primates in Japan.

Kinji Imanishi (1902–1992) was the leading figure during the dawn of Japanese primatology. Imanishi and his students—including Jun'ichirou Itani (1936–2001) and Masao Kawai (1934–)—launched the first study of Japanese monkeys in the wild in Koshima in 1948. In 1953 Koshima monkeys were found to have started a new practice: the washing of sweet potato, now a well-known example of "proto-cultural" behavior in monkeys. The study of Koshima monkeys still continues today and has so far recorded the history of a monkey community over eight generations.

After 10 years of accumulating knowledge on wild populations of monkeys, Imanishi and Itani first set out to East Africa in 1958 to begin the study of wild gorillas and chimpanzees. Thereafter, they sent a reconnaissance party to Africa every year and finally settled in Mahale in Tanzania in 1965. The study of Mahale chimpanzees, led by Toshisada Nishida, has now been progressing for more than 32 years and is the second longest longitudinal observation of a wild chimpanzee population.

Unique fieldwork focusing on monkeys and apes in the wild became the primary force in establishing KUPRI in 1967. KUPRI is the only national primate research institute in Japan, while there are, at present, eight national primate institutes in the USA. KUPRI has 40 faculty members, as well as numerous postdocs and graduate students—it is a research-oriented institution covering various (non-biomedical) disciplines. The institute sent Yukimaru Sugiyama to Guinea where he began the first longitudinal survey of wild chimpanzees (*Pan troglodytes verus*) in West Africa. The study at Bossou and the surrounding areas has now been running for more than 26 years, and still continues today.

I joined the study of wild chimpanzees at Bossou in 1986 as the second researcher following Sugiyama. Since then, I have visited Bossou and the surrounding areas in the Nimba mountains, a World Natural Heritage Site, every year. My research has focused on tool manufacture and use by the chimpanzees, an intellectual aspect of their life in the wild. Bossou chimpanzees are known to have unique cultural traditions of various kinds of tool-use. They use a pair of stones to crack open oil-palm nuts, leaves for drinking water, sticks for scooping algae floating on ponds, wands to fish ants, petioles of the oil-palm tree to practice pestle-pounding, and so forth.

To examine details of these tool-using behaviors, my colleagues and I began a field experiment.[21,22,23] We set up an outdoor laboratory where we provided nuts and stones in order to observe and video-record nut-cracking behavior in detail. We also drilled a hole in a tree trunk to provide fresh water that attracted the chimpanzees and encouraged their use of leaves for drinking water. We placed fresh oil palm nuts, dead caterpillars, and insects to attract safari ants— this provided us with the opportunity to observe ant-fishing behavior by the chimpanzees. In this way, we could watch and compare three kinds of tool-use at the same site.

This longitudinal field experiment has provided us with various interesting findings, such as perfect hand preference at the individual level and a weak shift toward using the right hand for stone hammering at the population level.

Chimpanzees have also been found to transport not only nuts but also stone tools, demonstrating a rudimentary form of possession of particular stone tools. The infants start using stone tools at the age of 4 to 5 years, which also marks the end of a critical period for learning. A form of observational learning, referred to as "Education by master-apprenticeship,"[21,22,23] plays a key role in the transmission of knowledge and skills from one generation to the next. This process is characterized by infants' prolonged exposure to the mother's tool-using behavior as a result of a close and long-lasting mother–infant bond, no formal instruction from the mother with either reward or punishment, the infants' intrinsic motivation to produce a copy of the mother's behavior, and high levels of tolerance by the mother toward their infants' activities during bouts of observation.

Our study of intelligence in the wild has been focusing on tool-using behaviors and has revealed the importance of social relationships in learning. Infant chimpanzees growing up in a particular community learn the unique cultural traditions of that community from the mother, the father, older siblings, and the other members of the group. We have been studying the processes underlying such intra-community transmission by introducing, in our outdoor laboratory, species of nuts unavailable at Bossou and therefore unfamiliar to the chimpanzees. Such manipulation has allowed us not only to track processes of cultural innovation and subsequent transmission within the group, but also to highlight inter-community transmission. Immigrant females will bring with them the knowledge acquired in their natal community, and through the spread of such knowledge within the group that they join, they contribute to the establishment of "cultural zones" in which neighboring communities come to share certain tool-using traditions, while remaining unique in their particular repertoire.

A NEW PARADIGM: STUDYING COGNITIVE DEVELOPMENT BASED ON A TRIADIC RELATIONSHIP

During the course of the Ai project, we succeeded in performing the first-ever artificial insemination of a chimpanzee in Japan. In March 1982, the first baby was born. Three more babies were born in 1982–83, two of whom were rejected by their biological mothers. I thus had an opportunity to raise a chimpanzee baby at home, and to compare the infant with my own. The experience led me to follow the classic study by Kohts, clearly revealing the similarities between the babies of the two species, humans and chimpanzees. At the same time, I recognized that comparisons of home-reared chimpanzees and home-reared humans are not fair because the chimpanzees are not reared by their own parents. I noticed that most of our knowledge on the cognitive development of infant chimpanzees came from "artificially" reared chimpanzees isolated from their conspecific community. Through my observations and field experiments in Africa, I truly recognized the importance of the community in which the infant chimpanzees are growing up.

Bearing this in mind, we devised a new paradigm for studying cognitive development in chimpanzees. It is based on a triadic relationship between a mother chimpanzee, an infant chimpanzee, and a human tester. The experiments take place in a large booth, which the experimenter enters with both the mother and her infant already inside. As the mother looks on, the tester presents the infant with a variety of tasks, and provides social reinforcement. In this way, the close bond established between the human experimenter and the mother—based on years of experience and daily interaction—allows us to test the infant chimpanzees in much the same context as that in which human infant developmental tests are conducted. In a face-to-face situation and with the mothers' cooperation, we are able to closely replicate many such tests, as well as design our own for illuminating developmental changes in the chimpanzee infants. We have been following the infants' development in object manipulation, the use of tools, drawing skills, the recognition of faces, facial gestural imitation, mirror self-recognition, the understanding of gaze and pointing as referents, and so forth. From such work and through our chimpanzee mother-chimpanzee infant-human tester triad paradigm we hope to establish one of the most accurate representations within the long-debated field of human-chimpanzee developmental parallels and contrasts.

REFERENCES

1. R. M. Yerkes and A. Yerkes, *The great apes* (Yale University Press, New Haven, 1929).
2. R. A. Gardner and B. T. Gardner, Teaching sign language to a chimpanzee. *Science*, **165**, 664–672 (1969).
3. D. Premack, On the assessment of language competence in the chimpanzee, in: *Behavior of Nonhuman Primates*, edited by A. M. Schrier and F. Stillnitz (Academic Press, NY, 1971), Vol. 4.

4. K. Murofushi, Numerical matching behavior by a chimpanzee (*Pan troglodytes*): Subitizing and analogue magnitude estimation. *Japan Psychological Research*, **39**, 140–153 (1977).

5. D. M. Rumbaugh, T. V. Gill, and E. C. von Glasersfeld, Reading and sentence completion by a chimpanzee (*Pan*). *Science*, **182**, 731–733 (1973).

6. D. M. Rumbaugh, *Language learning by a chimpanzee: The Lana Project* (Academic Press, NY, 1977).

7. T. Matsuzawa, Use of numbers in a chimpanzee. *Nature*, **315**, 57–59 (1985).

8. T. Asano, T. Kojima, T. Matsuzawa, K. Kubota, and K. Murofushi, Object and color naming in chimpanzees (*Pan troglodytes*). *Proceedings of the Japan Academy*, **59**(B), 118–122 (1982).

9. K. Fujita and T. Matsuzawa, Delayed figure reconstruction by a chimpanzee (*Pan troglodytes*) and humans (*Homo sapiens*). *Journal of Comparative Psychology*, **104**, 345–351 (1990).

10. N. Kawai and T. Matsuzawa, Numerical memory span in a chimpanzee. *Nature*, **403**, 39–40 (2000).

11. D. Biro and T. Matsuzawa, Numerical ordering in a chimpanzee (*Pan troglodytes*): Planning, executing, monitoring. *Journal of Comparative Psychology*, **111**(2), 159–173 (1999).

12. M. Tomonaga, Inversion effect in perception of human faces in a chimpanzee (*Pan troglodytes*). *Primates*, **40**, 417–438 (1999).

13. K. Hashiya and T. Kojima, Auditory-visual intermodal matching by a chimpanzee (*Pan troglodytes*). *Japanese Psychological Research*, **39**(3), 182–190 (1997).

14. J. Fagot and M. Tomonaga, Global and local processing in humans (*Homo sapiens*) and chimpanzees (*Pan troglodytes*): Use of a visual search task with compound stimuli. *Journal of Comparative Psychology*, **113**, 3–12 (1999).

15. N. Morimura, and T. Matsuzawa. Memory of movies by chimpanzees (Pan troglodytes) *Journal of Comparative Psychology*, **115**, 152–158 (2001).

16. K. Fujita, Perception of the Ponzo illusion by rhesus monkeys, chimpanzees, and humans: Similarity and difference in the three primate species. *Perception & Psychophysics*, **59**, 284–292 (1997).

17. M. Tomonaga, A search for search asymmetry in chimpanzees (*Pan troglodytes*). *Perceptual Motor Skills*, **73**, 1287–1295 (1993).

18. T. Matsuzawa, Spontaneous sorting in human and chimpanzee, in: *Language and intelligence in monkeys and apes: Comparative developmental perspectives*, edited by S. Parker and K. Gibson (Cambridge University Press, Cambridge, MA, 1990), pp. 451–468.

19. M. Tanaka, Object sorting in chimpanzees (*Pan troglodytes*): Classification based on physical identity, complementarity, and familiarity. *Journal of Comparative Psychology*, **109**, 151–161 (1995).

20. I. Iversen and T. Matsuzawa, Visually guided drawing in the chimpanzee (*Pan troglodytes*). *Japanese Psychological Research*, **38**(3), 126–135 (1996).
21. N. Inoue-Nakamura and T. Matsuzawa, Development of stone tool use by wild chimpanzees (*Pan troglodytes*). *Journal of Comparative Psychology*, **111**, 159–173 (1997).
22. T. Matsuzawa, Field experiments on use of stone tools in the wild, in: *Chimpanzee cultures*, edited by R. W. Wrangham, W. C. McGrew, F. B. M. de Waal, and P. G. Heltone (Harvard University Press, Cambridge, MA, 1994), pp. 351–370.
23. T. Matsuzawa, Primate foundations of human intelligence: A view of tool use in nonhuman primates and fossil hominids, in: *Primate origins of human cognition and behavior*, edited by T. Matsuzawa (Springer, Tokyo, 2001), pp. 557–574.

CHAPTER THIRTEEN

Language Studies with Bottlenosed Dolphins

Louis Herman conducts his dolphin research in tanks located at the Kewalo Basin Marine Mammal Laboratory in Honolulu, Hawaii. Visitors ascend stairs into a tower above two seawater tanks, one situated to the north and another to the south of the structure. The tanks are circular, 50 feet in diameter. The tanks are connected by a ten-foot-wide channel. Dolphins occupy both tanks.

As a visitor, my first impression was of the dolphins in the southern tank. My feeling was eerie, much as it was in a later visit to Pepperberg's parrot lab. The two dolphins in the tank swam lazily about, but there was nothing lazy about their examination of us human visitors: my wife, Dr. Terry Cronan, Dr. Vid Pecjak and his wife, Marinka, from Slovenia, and me. We all felt, rightly or wrongly, that we were both scrutinized and evaluated by the intelligences in the tank. Dr. Herman's research has demonstrated that dolphins can recognize individual humans, so it is reasonable to assume that they can make individual evaluations as well!

Dr. Herman quickly diverted our attention from this exercise in anthropomorphism by inviting us to witness the training in the northern tank. Trainers who are students at the University of Hawaii, or regular staff, were situated below us at tank level giving the dolphins instructions via large movements of the arms and hands, and rewarding them with fish for correct behaviors. Our crow's nest perch above gave us a bird's-eye view of the proceedings.

The dolphins in the northern tank could also watch television; that is, they could respond to visual commands given over a small television set mounted where the dolphins could see it behind an underwater window. The television cameras where the commands originated were mounted on tripods in the

213

command center where we stood, perhaps 20 feet above the tank. Dr. Herman then invited us to make movements that the dolphins would imitate. His assistants below had instructed a dolphin to attend to the TV screen. Dr. Herman opened the floor to anyone who wanted to do something that the dolphin might imitate. Dr. Pecjak, bold as always, went first, looking rather ridiculous standing on one leg while waving his other leg and arm in the air. The dolphin did its best—it came out of the water upright, bent its body, and waved its flipper in an ingenious portrayal of what Dr. Pecjak had just done. The rest of us then took a turn, bending forward, backward, and sideways by turns, in one direction or another, and accompanying our performances with different arm movements. In every case, the dolphin followed perfectly (within the limitations of its bodily structure). We were all flabbergasted at the dolphin's ability to understand the isomorphism between our bodies and hers.

The dolphins imitate humans who make large motor movements, like a pirouette, or obey other instructions if presented in their visual language. In formal tests, they imitate humans or other dolphins if given an "imitate" gesture, but otherwise do not, demonstrating that they do have a concept of the word.

One female dolphin, Akeakamai, has also learned to imitate new computer-generated sounds on the first try, and is thus the only animal to have demonstrated clear imitations of both vocal and motor models.[1]

Dr. Herman then showed us a broad array of sculptures made of plastic pipe, assembled in short pieces with a variety of connectors to form several very complex figures. He told us that the dolphins were very good at using their sonar to discriminate between the different pipe shapes when they were placed out of sight behind sound-permeable screens.[2,3] The dolphins could recognize the shapes "cross-modally"; that is, they could pick out an object sensed first acoustically by seeing it visually, and vice versa. Thus echolocation in the dolphin goes far beyond location, giving the dolphin enough information to recognize and match complex shapes initially seen either visually or acoustically.

Although we were all awed by the dolphin's remarkable abilities, they were not, of course, linguistic abilities, except for the ability to understand the signed commands. The dolphins' remarkable sensory ability, manifested in distinguishing complex shapes, is not much more remarkable per se than the bat's ability to catch insects in the dark, and we do not believe that bats are particularly intelligent. However, the ability to match cross-modally puts the dolphins' ability far beyond anything demonstrated by the bat.

Similarly, the ability to respond to commands like "imitate" may appear to be no more remarkable than a sheepdog's ability to herd sheep. However, the dolphin's understanding of the isomorphism between human and dolphin body structures that underlies this imitative ability is impressive, and imitating is a far more complex matter than is herding sheep. The sheepdog's behavior involves running off a series of learned behaviors, but imitation requires emitting behaviors that may be entirely new. Although all of these dolphin behaviors are complex and impressive, we must turn to additional experiments to demonstrate the details of dolphins' ability to comprehend language.

THE DOLPHIN BRAIN

Part of the human fascination with dolphins arises directly from observations of the intelligence suggested by their behaviors. Another important part, however, comes from studies of dolphin brains. There is wide variation in brain sizes among dolphin (odontocetes) species; however, some members of the delphinidae family with body sizes within the human range have brains as large as the human brain.[4] Body size is important in this connection because the encephalization quotient (EQ: ratio of brain weight to body weight) is correlated with measures of intelligence. *Tursiops truncatus*, one of the delphinidae, has an EQ roughly twice that of the higher primates; thus its EQ is between that of the great apes and that of humans. A measure related to EQ is the amount of "excess brain," brain that is not devoted to controlling bodily functions. The excess brain is then presumed to be available for intellectual functions—memory, perception, and problem solving, for example.

The dolphin brain is also very rich in fissures—that is, the cerebral cortex has multitudinous folds that increase the surface area of the brain, relative to its volume. It had been widely assumed that human brains had more folds than any other, but a study by Elias and Schwartz[5] indicated that an "index of folding" was 4.76 for a bottlenosed dolphin, and averaged only 2.86 for 20 humans. Thus it turns out that the most fissured brains known belong to the odontocetes. However, the dolphins' highly fissured cerebral cortex is much thinner than the human cortex, approximately half as thick. Calculations indicate that the total cortical volume is about 560 cubic centimeters, or about 80 percent of the human value of 660 cubic centimeters.

Although the dolphin brain is comparable in size to the human brain, it differs greatly in organization, far more than do ape brains.[6] One striking difference is that dolphin brains do not appear to be organized for binocular vision. Visual information from the two eyes is probably not combined; the dolphin's two eyes move independently and thus, unlike the apes' eyes, are probably not providing an integrated picture of the world. However, we should acknowledge that some researchers have suggested the possibility that dolphins might be able to increase their frontal vision by protruding their eyes to some extent. There may be some overlap in the visual fields at the nasal side, and the dolphin probably sees a broad continuous panoramic field encompassing at least 180 degrees. Thus dolphins rely heavily on vision as well as on sound.

Nevertheless, the two hemispheres of dolphins' brains are, visually speaking, probably largely independent. Physiologically, there is less connection between the two hemispheres than in humans. The main highway between the hemispheres, the corpus callosum, is relatively smaller in dolphins than in humans.

Hemispheric independence seems to carry over into dolphin sleep. Dolphins appear to sleep one hemisphere at a time. Such an arrangement may help an air-breathing animal that lives in water to maintain a reliable air supply while sleeping, and may be useful in avoiding predators.

The difference between humans and dolphins in the amount of brain devoted to auditory processing is even more striking than the difference in interhemispheric communication. Some parts of the dolphin brain are seven times as large as the homologous areas that humans devote to acoustic processing. Reliance on acoustic information is a useful dolphin adaptation to the marine environment. The brain areas devoted to acoustic processing are much closer to the areas devoted to visual processing than is the case for humans. In the dolphin brain, these areas are laid out in adjacent strips. Thus it seems likely that visual and acoustic information combine more intimately in dolphins' perceptual worlds than is the case for humans. As stated above, Herman found that dolphins could match complex objects seen only acoustically to objects seen visually, or vice versa. We could speculate that dolphins have bi-sensual vision rather than binocular vision!

There is little doubt that this array of differences between the dolphin brain and the human brain produces vast differences in the perceptual worlds of the two species.[7] Jerison suggests that the dolphin's mental world is organized more around audition than around vision. It would be interesting to observe dolphins' responses to an acoustic mirror—that is, a flat surface that reflects sound waves as effectively as a light mirror reflects light waves. So far as I know, such mirrors do not exist, but in principle one could reflect objects in the acoustic realm just as a mirror does in the visual realm. None of the above implies that dolphins are not also highly visual animals, as the cross-modal matching results so clearly demonstrate.

Because dolphins emit clicks and whistles that produce echoes in examining their acoustic world, some of the echoes in that world must flicker on and off. In a dolphin group that is emitting sounds it must be as though each member of a group of humans had a flickering flashlight that was turned on and off periodically to reveal the world around them. In that world the dolphin cannot see itself acoustically without an analog to a mirror or video camera. They can, of course, see themselves in light mirrors, and two dolphins have recently passed a version of Gallup's mirror test.[8] They did so by examining themselves in a mirror or reflective surface more quickly after real or sham marking, and attending to the real marks for a longer time than they attended to themselves when sham-marked.

Given the many differences between dolphins and humans, it is remarkable that any communication between them and us is possible. However, some communication—linguistic as well as nonverbal—is possible, so there must be substantial overlap between our worlds, in spite of the differences. Perhaps the Austrian philosopher Wittgenstein was wrong when he claimed that if a lion could talk, we wouldn't understand him. The lion's world could hardly differ more than the dolphins from ours.

AVOIDING THE ERRORS OF EARLIER RESEARCHERS

Despite the differences in organization, the size and surface area of the dolphin brain made dolphins promising candidates for acquiring rudimentary human language. Herman, Richards, and Wolz[9] wrote the first full report of efforts to teach bottlenosed dolphins to comprehend an artificial gestural and an artificial acoustic language. Their report starts with a balanced review of ape language research. Herman et al. were aware of Terrace's criticisms, and of responses to them, and they wanted to avoid the errors of the past.

Herman et al. called for better approaches that would not be subject to the earlier criticisms, and advocated an emphasis on comprehension rather than on production. They argue that a study of comprehension offers better control and more objectivity, citing McNeill, who said " ... in comprehension the investigator knows what the input to the process is—it is the sentence comprehended. Thus when comprehension fails, the source of trouble can be located. The same cannot be said for production (p. 209). Lilly's work is cited critically, and Batteau's start toward teaching dolphins an artificial language is acknowledged.

Herman et al. made the following thought-provoking complaint against Batteau:

> Individual whistles did not refer to individual semantic entities, i.e., to unique words, but instead merely set the occasion for a chain of responses. For example, a single, particular whistle sound instructed the dolphin to "hit the ball with your pectoral fin." There were no separate words for "hit" or "ball" or "pectoral fin."

This raises the issue of what is allowable as a semantic entity. Is it not legitimate to have one entity that means "let's go upstairs to bed and make love?" Of course a mere glance can convey that message, but that is precisely the point Herman et al. were making. A glance or multi-word grouping is not accepted as a component of language because neither combines with other linguistic elements to construct complex utterances. Neither, they believe, can the type of whistles used by Batteau.

DOLPHIN THEFT: DEATH AND REBIRTH OF A PROJECT

Animal activists stole and released two partly language-trained dolphins with whom the researchers in Herman's laboratory had been working. One, Keakiko (Kea for short) had been taught three symbols for objects: for a ball, a Styrofoam cylinder, and a life ring, and three symbols for actions: to fetch, to touch, and to mouth. The training used with Kea, and later with other dolphins, involved pairing signs with objects and gradual shaping of the dolphin to perform specified actions with individual objects in response to signs.

Kea responded correctly to new instances of each type of object, and "Under these conditions Kea quickly came to respond nearly flawlessly to each of the nine possible two-word sentences generated by the six vocabulary items (for example, CYLINDER MOUTH, RING TOUCH, etc.)." The study ended abruptly in May of 1977 with the abduction of Kea from the laboratory. In a footnote, Herman expresses bitterness[9] "The pair was transported in a small van to a remote location about 50 miles from their home tank and abandoned in the open ocean. The sudden demands of open-ocean living combined with the stress of removal from familiar surroundings posed a cruel survival test for these long-time domesticated dolphins that surely could not be met" (p. 135).

Kea and Puka, the other dolphin that was abducted, were replaced after 2 years by two young females, Akeakamai (Ake) and Phoenix, caught near Gulfport, Mississippi, in June 1978. Phoenix's name was chosen to symbolize the rebirth of the lab. They were thought to be 2 to 3 years old.

HERMAN DEVELOPS DOLPHIN LANGUAGE TRAINING TECHNIQUES

To prepare the dolphins for their training, Herman and the others socialized with them by swimming and playing with them daily in their tank, stroking them, interacting with them, showing them things through the windows of their tank, and by feeding them. Socialization training was started within a very few days after their collection in Mississippi Sound.

After habituation and socialization, Ake and Phoenix were trained on two-choice discriminations with freshly thawed silver smelt and extensive social praise as rewards for correct responses, to give them the idea that problems could be solved and to train them on learning sets. Although fish were used as a reward, the dolphins were never deprived of food. Most of the discriminations were acoustic, but the dolphins responded so well to gestures that Ake was transferred to visual language training, while Phoenix was trained on the acoustic language as originally planned.

The dolphins' formal language training started within the first several months. Several techniques were used during this phase. Most involved direct instruction—pointing to the correct item or using other types of prompts. Shaping was seldom used, but humans modeled the correct responses, and modeling is of course an integral part of imitation studies.

Modeling was used more often in later stages of training after the dolphins had developed some sophistication. Pointing was used to direct behavior, and objects were touched to the dolphin's body to indicate the part that the dolphin was to use in touching the object. The dolphins were themselves sometimes moved to a location or position. To teach concepts like "over," objects were held under water in the beginning, and the dolphin was directed to swim over it in the presence of the OVER gesture; a target pole was used to direct the dolphin. Then the object was gradually raised until the dolphin had

Dr. Herman tells Akekamai to pay attention; note the large scale of the sign, made by using both arms and hands. Photo courtesy of Dr. Louis Herman.

to jump over it. IN was taught by placing an object in the dolphin's mouth and then encouraging her to place it in an adjacent basket in the presence of the IN sign. Analogous techniques were used for other signs.

Phoenix's acoustic language was "mainly whistle-like." Dolphins produce whistles naturally in the wild. However, the authors tried to avoid the natural whistles, no doubt to try to reduce confusion. The exception was that the most commonly used individual whistle of an individual dolphin was assumed to be its signature to other dolphins, and was used as its name.

In the acoustic language, whistle-words were presented via a computer keyboard that controlled waveform generators that played through underwater speakers. Frequencies could go to over 40 kHz, but most whistles were at least partially within the human range that stops at around 20 kHz. Herman et al. present the waveforms for the different words; when Phoenix had trouble, the sound was lengthened or changed to make it easier for her. Dolphins hear best between 15 and 90 kHz, although their hearing extends as high as 120 kHz and as low as 100 Hz.

The gestures used for Ake's language were (to humans) easily visible, "moderate to large-scale" movements of the hand(s) and arm(s). Previous work indicated that dolphins have great difficulty working with static displays, so easily visible movements were incorporated into the gestures. Objects and object modifiers were indicated by moving both arms, and actions were denoted by moving only one arm.

RULES FOR DOLPHINS' ACOUSTIC AND GESTURAL LANGUAGES

The acoustic and gestural languages had similar vocabularies and rules for sentence formation, with some exceptions. Phoenix's rules were generally more like those used in English. In both languages, action symbols in two-word sentences come last; for example, HOOP TAIL-TOUCH means "Go to the hoop and touch it with your tail." For Phoenix, SURFBOARD FETCH BALL means "Take the surfboard to the ball," so the action, like the sentence as glossed, proceeds from the left to the right; the progression therefore is from the first sign presented to the last sign presented. For Phoenix, the meaning of the symmetrically reversed sentence, BALL FETCH SURFBOARD, is itself symmetrical; that is, it would mean "Take the ball to the surfboard." For Akeakamai however, "Take the ball to the surfboard" would be conveyed by SURFBOARD

BALL FETCH, with the action term always coming last, the "goal" object first, and the item to be transported second. In Ake's language, the meaning of sentences relating two objects is reversed by reversing the positions of the two object signs, from, for example, BALL SURFBOARD FETCH to SURFBOARD BALL FETCH. This reversibility allowed the researchers to determine whether the dolphins were sensitive to the syntactic rules governing such sentences, and both dolphins showed that they were very aware of this feature of their language. In addition, it seemed to matter little which of the two rules was applied to generating sentences; Ake and Phoenix did about equally well overall, although both the rules for the placement of the direct and indirect objects and the placement of the action word were reversed.

DOLPHIN PERFORMANCES WITH THEIR ARTIFICIAL LANGUAGES

Their ability to respond correctly to the imperative sentences on which they were trained was impressive. Herman et al. were careful to control for nonlinguistic cues and to use novel sentences to assess the dolphins' comprehension. An observer who did not know what instruction the dolphin had been given wrote down his or her description of each response in the appropriate artificial language. For example, if Phoenix heard the instruction FRISBEE FETCH, and instead fetched the hoop, the observer would write HOOP FETCH, and the response would be scored as an error. Overall, the dolphins were about 75 percent completely correct in responding to novel commands of this type. The probability of making a correct response randomly was usually from 1 percent to 2 percent, and never as much as 4 percent, given the number of available items from which the correct item or items had to be chosen and the number of responses from which the correct response had to be chosen.

Herman et al. argued that their results were the best evidence up to that time for the ability of any non-human animal to respond to syntactic information. They found it noteworthy that Ake and Phoenix did about equally well with two different grammars; that argues against any nativistic preference for one type of ordering over the other.

Herman[10] emphasized almost from the beginning of his research the distinction between two types of sentences, a point that he reiterated forcefully in his response to some criticisms by Schusterman and Gisiner.[11] A "non-relational" sentence is of the type FRISBEE FETCH. Such sentences involve only a single object and a single action, and dolphins are quite good at responding to such sentences, as noted above. The relational sentences, several of which were cited above, involve an action that relates two objects, as in BALL FRISBEE FETCH. This would be glossed in English as "get the frisbee and take it to the ball," in Ake's visual sign language.

The dolphins made more errors on relational sentences than on non-relational sentences. In work prior to 1980[9] Phoenix was about 50 percent correct on relational sentences of the type BALL FETCH FRISBEE. Most of Phoenix's errors were to the second, goal, object; that is, in the above sentence Phoenix was very likely to fetch the ball, but not as likely to fetch it to the frisbee. Errors were most frequent when they were told to relate one trans-portable object to another; for example, BALL FRISBEE FETCH in Ake's language was more likely to produce an error than BALL WATER FETCH because the posi-tion of the water hose could not be changed by the dolphin, but both ball and frisbee could be anywhere in the tank. Both dolphins, in spite of the fact that the indirect object sign came first for Ake and last for Phoenix, were more likely to mistake the destination object (the indirect object in the sentence) than to mistake the object to be moved (the direct object in the sentence). These results might have occurred because of the additional search time required to find the second object, which could have caused forgetting what that object was—especially when the object was movable.

Although the ability of Ake and Phoenix to obey signed commands is impressive, the true test of language comprehension must come from responses to novel sentences. The dolphin responses to lexically novel sentences ranged from 56.0 to 86.7 percent correct. (A lexically novel sentence is one in which a particular sign has never appeared before in a par-ticular sentence slot.) This is an amazing performance, when one considers the complexity of some of the novel sentence types tested. Some sentence types were simple, consisting only of a direct object and an action, for example, PIPE FETCH. But others were quite complex, consisting of the string modifier-direct object-relation-modifier-indirect object. On this complex type of novel sentence, Phoenix was correct on 18 of 30 test commands, with the more frequent errors occurring on the indirect object or its modifier, as expected from her pattern of errors on sentences that were not novel.

The dolphins were also tested on sentences that included new relational terms (like IN) that had never been used in sentences before. Akeakamai had been trained to the sign IN in isolation; it directed her to place any object of her choice in another object. When it was added to a sentence BASKET HOOP IN, she immediately swam to the hoop, carried it to the basket, and dropped it in. She responded wholly correctly to 66 percent of 128 relational sentences presented to her.[1]

When tested on sentences whose meaning was changed because of a change in word order, both did well (Akeakamai, 59 percent; Phoenix, 77 percent). They also performed reasonably well when tested on conjoined sentences, although sometimes, as might be expected, they performed only one of the actions indicated, probably because the longer sentences challenged their memories. The total picture indicates that dolphin performance exceeds what can reasonably be accounted for in terms of conditional discrimination learning, and indicates a degree of syntactic understanding.

Herman and Uyeyama[12] also believe that dolphins have demonstrated some understanding of closed class words (prepositions, conjunctions, demonstratives, and locatives), although Kako[13] had raised questions about whether existing animal language research had been sufficient to prove the capacity to acquire this understanding. In the same connection, Herman and Uyeyama argue that the dolphins' responses to anomalous sentences (details are presented below) demonstrated some understanding of "argument structure."

DOLPHINS KNOW THEIR BODY PARTS

Finally, every parent can appreciate the last feat demonstrated by Herman's team and the dolphin, Elele.[14] Hardly a parent lives who has not taught his or her child to "Point to your nose," to respond to "Where's your eye?", and so on. Herman et al. taught Elele the gestures representing nine body parts, and showed that she understood which part was represented by the gesture. Elele was tested by requesting that a specified body part be used to touch a specified object in the pool. Elele also learned to display the referenced body part or shake it back and forth upon request. Performance on all the tasks was far better than chance, and in many cases Elele used the specified body part successfully when a new use was requested. A human parent would have praised such a smart child profusely!

DOLPHINS UNDERSTAND POINTING

Language-trained apes typically learn to use pointing as an indicator of a person who is to carry out an action. Panzee learned to point to locations where food was hidden. Paradoxically, and in contrast to the fact that children and chimpanzees alike usually can comprehend words before they can produce them, chimpanzees find it difficult or impossible to interpret pointing by humans.

Dolphins present at least as great a paradox with respect to pointing.[15] They do understand human pointing. The researchers put objects to the left, right, or behind the dolphin, and the dolphin chooses the object to which the researcher points. For example, in response to pointing at the hoop, followed by the gesture UNDER, the dolphin swam under the hoop. The dolphin could even respond correctly to two points followed by an action gesture, for example: point to basket, point to hoop, gesture FETCH, and the dolphin took the hoop to the basket. The ability to interpret points is especially surprising because dolphins have no arms and do not use gestural pointing with one another. Perhaps they are used to acoustic pointing, and perhaps they attend especially well to human arms because their words are conveyed via arm movements.

Thus dolphins *comprehend* the referential function of pointing, the other side of the apes' understanding of the referential function of the pointing they *produce*.

EVALUATING THE DOLPHIN RESEARCH

Schusterman and Gisiner,[11] following Premack in this respect, objected to some of Herman's nomenclature and several of his conclusions. Schusterman and Gisiner quote Premack as saying that Herman's "flurry of linguistic terms is gratuitous." For example, the former authors complained about Herman's use of grammatical terms like the above-mentioned direct and indirect objects, and added parse, grammar, sentence, lexical, and "semantic proposition" as additional unnecessarily linguistic terms.

The present authors agree with Premack, Schusterman, and Gisiner in being quite certain that dolphins have no concepts corresponding to the grammatical notions of direct and indirect objects, etc., and Herman would no doubt agree. However, we, like Herman, see no harm in his using the terms as an economical way of describing his use of signs to other humans who do have these grammatical concepts. There are less suggestive ways of describing the structure of the language presented to the dolphins; however, what these other ways of talking gain in theoretical parsimony they may well lose in descriptive economy. We think readers bear as much responsibility for limiting what Herman means as he does for limiting the possibilities in his writing. It should be clear from the context that Herman intends only to describe what he presents to the dolphin, not what the dolphin knows about what he is presenting. Thus Herman and the reader know what a direct object is in general, but the dolphin presumably knows only what position a particular sign occupies in a sequence of from two to five signs.

What Premack, Schusterman, and Gisiner prefer is a language more closely related to traditional experimentation in psychology. It is always possible to give different explanations of any observed behavior, including behaviors that are language-like. Thus terms like "higher order conditioning," "stimulus equivalence," and "conditional discrimination" are preferred by some because they are alleged to refer to simpler processes than "parsing" or identifying the indirect object. However, Herman[13] notes that Premack and Schusterman are themselves not too pure in their avoidance of linguistic terms:

> Premack, in his major treatise on the language work with the chimpanzee Sarah, liberally sprinkled linguistic terms throughout that work, not only terms like "sentence" and "word," but more ponderous ones like "demonstrative pronoun," "demonstrative adjective," "nominative phrase," and "accusative phrase". (Premack, 1976, pp. 320–321)

Perhaps linguistic terms are gratuitous only when used by animal language investigators other than oneself! As we learn more about animal competence, we will be better able to decide which theory and associated set of terms renders the more accurate account of observations, and to develop better rules for deciding when each type of vocabulary is appropriate. Herman, Premack, and Schusterman are by no means the first to debate this rather old issue, and they are not destined to be the last.

Schusterman and Gisiner also complained about Herman's calculation of the probability that a dolphin would respond correctly by chance. There are indeed several ways of computing such probabilities. The Gardners[17] give a nice overview (pp. 334 ff.) of some aspects of this problem (naturally enough; they worked in Reno, one of our gambling meccas, where probabilities are critical). If a command consists of an object plus an action, for example, FRISBEE OVER (jump over the frisbee), an obvious procedure would be to look at the number of objects in the pool and the number of responses available to the dolphin. If there are eight objects, the probability of acting on the correct one is $\frac{1}{8}$ if the dolphin is unbiased. If there are five actions, the probability of choosing the correct action is $\frac{1}{5}$. The probability of making a completely correct response is then $\frac{1}{8} \times \frac{1}{5}$, or $\frac{1}{40}$. The same conclusion is reached if we calculate the number of things that can happen—five actions could be taken on each of eight objects, so there are 40 possible things that can happen. If each is equally likely, then the probability that the one correct combination would be chosen out of 40 is again $\frac{1}{40}$.

It is worth noting, however, that these seemingly simple calculations are based on a number of assumptions. For one thing, the dolphins probably prefer some objects and some actions to other objects and other actions. So some commands are more likely than other commands to elicit correct responses. That should not affect the overall probability of being correct, but it does mean that the experimenter must be careful to present a representative sample of commands. Another consideration is that the whole calculation is based on the assumption that only the specified objects can be selected, and only the specified actions can be taken. To see why this is so, ask yourself this question: How likely is it that an untrained dolphin, in the pool for the first time, would, upon seeing the signs FRISBEE HOOP FETCH, take the hoop to the frisbee? We would all agree that the probability is close to zero. Finally, this calculation assumes that the probabilities have not been modified by prior trials in some fixed sequence of trials.

Any assignment of probabilities thus rests upon assumptions that are often implicit and unexamined. In an animal language study, the usual concern is with how exposure to a symbol or sequence of symbols changes the response probabilities obtained prior to the observation of the symbol or symbols. If the probabilities change markedly, we say that the recipient understood the message to some degree. The initial probabilities, prior to the receipt of a particular message, thus depend on the stage of training at which we find the organism. A tremendous amount of preparation is necessary before we can say that the dolphin, or other organism, is going to select only from a specified set of responses; the untrained dolphin is more likely to do a roll, or jump out of the water, than to fetch something. Even the trained animal may not make a response that can be scored, so the experimenter then records a balk or does not count the trial, or recognizes that more training is necessary before testing can begin.

Thus assignment of probabilities requires setting up an abstract model of what random responding would be in a given situation. Herman used the type of model exemplified above for assigning probabilities to correct responses to relational command sequences. For example, if there are eight relevant objects and five possible responses, then (assuming that all combinations are equally likely) the probability that BALL FRISBEE FETCH would elicit a correct response by chance is simply $\frac{1}{8} \times \frac{1}{7} \times \frac{1}{5} = \frac{1}{280}$ or about $\frac{1}{3}$ percent. Ake almost always did better than that, even in response to command sequences that she had never seen before, and often significantly better.

Schusterman and Gisiner[9] preferred another method of assigning probabilities to correct responses. Noting that Herman's dolphins (and their own seals) almost always chose the correct action and direct object, they suggested that the probability of correct response should depend only on the number of objects in the pool—and since one had already been chosen as the object to be transported, the remaining choices were one less. Thus, if there were eight objects in the pool, Schusterman and Gisiner suggested that the chance probability of a correct response to the above command should be $\frac{1}{7}$!

This is an unusual claim. However, note that it is appropriate for answering a much more limited question than the question answered when we ask whether the dolphins, or seals, or chimpanzees, or whatever, can respond correctly significantly more often than once in 280 tries. Schusterman and Gisiner are *assuming* that the dolphin understands the action and the direct object perfectly, and then asking only whether the animal, in this context, does better than chance in selecting the indirect object. This is a reasonable question to ask, but it is a quite different question from "Is the animal responding at better than chance to this command?" The latter question is the one being answered when we use Herman's probability assignment. Thus the real issue with respect to assigning probabilities is not "Which assignment of probabilities is correct?" but rather "What assumptions are you making?" The answer to the latter question, in turn, determines what question is being asked and answered.

Herman replied[13] to Schusterman and Gisiner's commentary as follows: "Schusterman and Gisiner calculate chance probabilities for responding to goal objects as the number of objects in the tank minus one (the object that is being transported is not counted). They criticized our method of calculating chance probabilities which, however, was for *whole* sentences, not goal objects. We used a finite-state model in which it was assumed that the syntactic form of a sentence frame constrained the responses the dolphin might make."

Herman's response is exactly in keeping with our comments above. We should note, however, that it may be interesting to ask both types of questions, the one about complete sentences and the one about indirect objects. The latter question is of particular interest because the dolphins' ability to respond to serial position cues depends largely on their ability to respond at above chance levels to the goal object. If Ake, for example, always responded correctly to the direct object, which immediately precedes the action word, and never to the indirect object, which occurred earlier, the performance could be attributed simply to forgetting the first-occurring object. That would not be a very impressive demonstration of sensitivity to serial position and thus not a very convincing demonstration that dolphins have any ability to analyze syntax.

However, if dolphins *generally* transport the second object in the sentence to the correct first object, even if they do not always do so, and never transport the first object to the location of the second, then we know that they are sensitive to serial position, not just forgetful. Phoenix's behavior provides additional evidence that simple forgetting cannot be the correct explanation, since in her case the direct object preceded the indirect object, but most errors still involved the indirect (goal) object. Retroactive inhibition (forgetting the first object while obtaining and transporting the second) could, however, account for her behavior, and in Ake's case both the early position in the sentence and the need to act on the direct object first could lead to forgetting.

The bonobo Kanzi displayed a systematic pattern of errors on novel sentences that seems to have features in common with the dolphins' errors. Kanzi did better than the child, Alia, with whom he was compared, on sentences containing embedded phrases, like "Go get the ball *that's outside*." But Alia did better than Kanzi on conjunctive sentences, like "Go get the ball and the doggie." On such sentences, Kanzi was very likely to fetch the ball but not the doggie, thus displaying something of a one-track mind. When dolphins find a transportable object, but take it to the wrong place, they exhibit a similar fixation on a part of the task. Further, when dolphins are given anomalous sequences of signs[18] they often respond to a subset of the sequence that conveys a normal message. For example, Ake behaved as though she extracted normal subsets from 17 of 19 "illegal" sentences that contained such subsets. A sample anomalous sequence is SPEAKER HOOP PIPE IN, which illegally contains three objects, rather than two, before the action IN. Legal subsets, without changing serial order, are then SPEAKER HOOP IN, SPEAKER PIPE IN, and HOOP PIPE IN. Ake typically chose one legal action, for example by placing the hoop on top of the speaker. It appears that an animal is likely to proceed directly to a single action when one can be found, ignoring either conjunction or anomaly, depending on the context. However, when there was no legal action specified within an anomalous sequence of signs, Ake often refused to respond, which she never did when some embedded sequence was legal.

Even if the Schusterman and Gisiner method is used to calculate probabilities, the dolphins performed above chance levels even on the most difficult of their familiar sentences. However, as of 1989, Ake, who at that time performed better than Phoenix, had not always performed above chance levels on all types of novel sentences. For one type, Ake had been tested with only one novel sentence, and for another type with none. In a third test of a type of novel sentence, indirect object, direct object, action (for example, BALL FRISBEE FETCH), six of seven responses were errors, all because the wrong indirect object was chosen. For a fourth type of novel sentence, which on the surface might appear more difficult (LEFT BALL FRISBEE FETCH would be an example), Ake got 17 of 24 commands perfectly correct! Perhaps the presence of a second cue to the correct destination (LEFT, in addition to BALL) helped Ake to remember where she was going. Consistent with that reasoning, Ake's performance was more reliable when the goal object was not transportable than when it was transportable.

One wants to know, of course, what is different about the performances of these language-trained dolphins and the very clever performances of the dolphins seen in shows at Sea World and similar venues. This issue is easily settled; dolphins in shows are trained to perform specific acts in response to specific cues. The cues cannot be recombined to produce new action sequences. Akeakamai and Phoenix interpreted new sequences of symbols correctly and performed corresponding new sequences of actions with no specific training on the new sequences. That is the kind of ability that distinguishes animals who understand the separate symbols in a language from those who have merely learned to do tricks in response to cues.

DO SIGNS REFER FOR DOLPHINS?

It is always difficult to prove that spoken words, written words, signs, lexigrams, plastic symbols, or any other symbols intended to represent objects, actions, or relationships in the real world really refer. An alternative explanation is that the supposed symbol is simply a discriminative stimulus to which the animal (including the human animal) has learned to respond. Even when the response to the same symbol is different in different contexts (including the context of accompanying symbols) the skeptic can claim that the animal has learned a conditional discrimination; another way of putting it is that the animal treats the totality of stimulation (symbol plus accompanying symbols or other context) as a single complex stimulus to which the appropriate response can be made.

Schusterman and Gisiner are exponents of the conditional discrimination account of dolphin learning. They support their criticism of the "top-down," linguistic, explanation of dolphin capability by replicating much of the dolphin work with sea lions (*Zalophus californianus*).[18] These authors took a "bottom-up" approach involving five stages to training their sea lion, Rocky. Rocky's ability to respond to gestures was similar to that of the dolphins.

In the first stage, Rocky was taught to take an action in response to a gestural sign. In the second stage, he was taught to respond to different objects in response to different signs. The third stage combined the first two; at this point the discrimination becomes conditional, because the action taken depends on which object was indicated by the accompanying gesture; that is, the response OVER in PIPE OVER is conditional upon which gesture (in this case it indicated PIPE) preceded it. In the fourth stage of training the conditional cues were made more complex by adding size and brightness modifiers before the gestures for objects. For example, the conditional cues could be SMALL WHITE preceding CONE. The fifth stage involved training Rocky to perform relational actions, for example, taking one object to another. In this final stage a complete command might be WHITE LARGE CUBE, SMALL BLACK BALL FETCH.[19] At this stage the conditional cues for cube are "white" and "large," for ball they are "small" and "black," and the conditional cues for fetch are all the preceding gestures.

The explanatory burden of this type of explanation becomes very heavy when the number of complex stimuli is large. It becomes too heavy to bear when an animal responds to a new combination of stimuli with the correct response, with no training to do so. However, Schusterman and Gisiner argue that animals can succeed in this context because objects that occupy the appropriate position in a sentence frame are "functionally equivalent." That is, the appropriate action may occur to CONE if it is substituted in the position that has previously been occupied only by BALL and RING. However, saying that the different gestures are functionally equivalent smacks of circularity (the items are functionally equivalent if the animals respond in the same way to them, and vice versa).

One also wonders how items become functionally equivalent if they do not symbolize a referent. The account in terms of functional equivalence may, therefore, provide less explanatory power than admitting that the gestures refer to the objects and that the dolphin has some understanding of syntax.

An intermediate stage of explanatory difficulty occurs when a single symbol is responded to differently in many contexts. A good example from Herman's work with Akekamai is her multiple responses to the single sign PERSON.[20] The PERSON sign occurs in six types of gesture strings. Examples are PERSON UNDER, a nonrelational imperative telling Ake to swim under the person in the tank; other action gestures can also be used in this context. PERSON HOOP FETCH is a relational imperative asking Ake to take a hoop to the person, with its opposite, HOOP PERSON FETCH, asking Ake to take the person to the hoop. Again, other action gestures can be used in this context. PERSON can also be used as a simple interrogative, asking Ake to indicate by pressing a YES or NO paddle whether a person is present in the tank. Similarly, three other classes of gesture strings can include PERSON, and the combination of all types of gesture strings with all types of requested actions yields a bewildering number of combinations into which PERSON can enter.

It is still possible to maintain that Ake could have learned conditional discriminations that guided her to the correct response for every string in which PERSON occurred. However, the explanatory load appears to be lighter if we simply admit that Ake understands that PERSON refers to a human being, and that she can use that information to decode a variety of commands relating to humans. This line of thought supports the argument that a greater than chance number of correct responses to novel gesture strings indicates some degree of dolphin understanding of the referential function of gestures and of the significance of the order in which the gestures occur.

Ake gave one additional indication that she recognized the referential use of her symbols. She reliably reported whether objects were present in or absent from her tank.[21] For example, a frisbee and a ball would be placed in her tank, and she would be asked whether there was a frisbee in the tank; she reliably indicated its presence by pressing a paddle to her right. If she were queried about a hoop under this condition, she reliably pressed a paddle to her left, indicating that she understood the gesture HOOP even though no hoop was present.

ANOTHER INDICATION OF DOLPHIN SELF-AWARENESS

In addition to passing the mirror test, dolphins have indicated awareness of their own recent behaviors by responding correctly to orders to REPEAT their last performance or to perform ANY other of a set of five actions.[22] For example they might be told to TAIL TOUCH an object, followed by REPEAT, whereupon they repeat the tail touch. Or the tail touch might be followed by ANY, whereupon they could choose among the remaining actions, OVER, UNDER, MOUTH, or FLIPPER TOUCH. Although the mirror test has been the gold standard of self-awareness, awareness of past actions appears to be an equally justified alternative measure.

LEARNING LANGUAGE: DOLPHIN AND CHILD

Sue Savage-Rumbaugh[23] pointed out four differences between the situations in which dolphins must learn language and the situations in which human children learn language. The human, but not the dolphin, situation makes it possible to acquire language through observation.

Unfortunately, for the dolphin there is neither the motivation nor the ability to invert the linguistic relationship between the experimenter and the dolphin, as there is in the case of apes, who can use lexigrams, sign language, or plastic symbols (depending on the technique being used to train them) to become the speaker rather than the listener.

There are, therefore, serious difficulties involved in working with animals who are very different from humans and live in an environment so foreign to us. However, some of these difficulties may be circumvented; there have been several attempts to provide dolphins with response boards similar to lexigram boards. No publications have so far reported much success in these efforts, although one study[24] did indicate that the two male offspring of two female dolphins spontaneously used six underwater lexigrams to obtain objects or activities. Pressing the lexigrams also produced associated sounds that the male dolphins learned to mimic. The females, who obviously were older when the lexigrams were introduced, did not use the board.

However, given dolphins' remarkable behavior in the face of difficulties, we may look forward to much more remarkable performances by dolphins in the future. One way to keep up with that future is by consulting Dr. Herman's web site at http://www.dolphin-institute.org/.

REFERENCES

1. L. Herman, Cognition and language competencies of bottlenosed dolphins, in: *Dolphin cognition and behavior: A comparative approach*, edited by R. J. Schusterman, J. A. Thomas, and F. G. Wood (Erlbaum, Hillsdale, NJ, 1986).

2. L. Herman, A. Pack, and M. Hoffman-Kuhnt, Seeing through sound: Dolphins (*Tursiops truncatus*) perceive the spatial structure of objects through echolocation. *Journal of comparative psychology*, **112**(3), 292–305 (1998).

3. A. A. Pack, L. M. Herman, and Hoffmann-Kuhnt, Dolphin echolocation shape perception: From sound to object, in: *Echolocation in bats and dolphins*, edited by J. Thomas, C. Moss, and M. Vater (University of Chicago Press, Chicago, in press).

4. S. M. Ridgway, Physiological observations on dolphin brains, in: *Dolphin cognition and behavior: A comparative approach*, edited by R. J. Schusterman, J. M. Thomas, and F. G. Wood (Erlbaum, Hillsdale, NJ, 1986), pp. 31–59.

5. H. Elias and D. Schwartz, Surface areas of the cerebral cortex of mammals determined by stereological methods. *Science*, **166**, 111–113 (1969).

6. L. Marino, K. Sudheimer, T. Murphy, K. Davis, D. Pabst, W. McLellan, J. Rilling, and J. Johnson, Anatomy and three-dimensional reconstructions of the brain of a bottlenose dolphin (*Tursiops truncatus*) from magnetic resonance images. *The Anatomical Record*, **264**, 397–414 (2001).

7. Harry J. Jerison, The perceptual worlds of dolphins, in: *Dolphin cognition and behavior: A comparative approach*, edited by R. J. Schusterman, J. A. Thomas, and F. G. Wood (Erlbaum, Hillsdale, NJ, 1986), pp. 141–166.

8. D. Reiss and L. Marino, Mirror self-recognition in the bottlenose dolphin: A case of cognitive convergence. *Proceedings of the National Academy of Sciences*, **98**(10), 5937–5942 (2001).

9. L. M. Herman, D. G. Richards, and J. P. Wolz, Comprehension of sentences by bottlenosed dolphins. *Cognition*, **16**, 129–219 (1984).

10. L. M. Herman, In which Procrustean bed does the sea lion sleep tonight? *The Psychological Record*, **39**, 19–50 (1989).

11. R. J. Schusterman and R. Gisiner, Artificial language comprehension in dolphins and sea lions: The essential cognitive skills. *The Psychological Record*, **38**, 311–348 (1988).

12. L. M. Herman and R. K. Uyeyama, The dolphin's grammatical competency: Comments on Kako. *Animal Learning and Behavior*, **27**(1), 18–23 (1999).

13. E. Kako, Elements of syntax in the systems of three language-trained animals. *Animal Learning and Behavior*, **27**(1), 1–14 (1999).

14. L. Herman, D. Matus, E. Herman, M. Ivancid, and A. Pack, The bottlenosed dolphin's (*Tursiops truncatus*) understanding of gestures as symbolic representations of its body parts. *Animal learning and behavior*, **29**(3), 250–264 (2001).

15. L. Herman, S. Abichandani, A. Elhajj, E. Herman, J. Sanchez, and A. Pack, Dolphins (*Tursiops truncatus*) comprehend the referential character of the human pointing gesture. *Journal of comparative psychology*, **113**, 1–18 (1999).

16. L. M. Herman, The language of animal language research: Reply to Schusterman and Gisiner. *The Psychological Record*, **38**, 349–362 (1988).

17. R. A. Gardner and B. T. Gardner, *The structure of learning. From sign stimuli to sign language* (Erlbaum, Mahwah, NJ, 1998).

18. L. M. Herman, S. A. Kuczaj, II, and M. D. Holder, Responses to anomalous gestural sequences by a language-trained dolphin: Evidence for processing of semantic relations and syntactic information. *Journal of Experimental Psychology: General*, **122**(2), 184–194 (1993).

19. R. J. Schusterman and R. Gisiner, Pinnipeds, porpoises, and parsimony: Animal language research viewed from a bottom-up perspective, in: *Anthropomorphism, anecdotes, and animals*, edited by R. W. Mitchell, N. S. Thompson, and H. L. Miles (State University of New York Press, Albany, 1997), pp. 370–382.

20. L. M. Herman, A. A Pack, and P. Morrel-Samuels, Representational and conceptual skills of dolphins, in: *Language and communication: Comparative Perspectives*, edited by H. L. Roitblat, L. M. Herman, and P. E. Nachtigall (Erlbaum, Hillsdale, NJ, 1993), pp. 403–442.

21. L. Herman and P. Forestell, Reporting presence or absence of named objects by a language-trained dolphin. *Neuroscience and Biobehavioral Reviews*, **9**, 667–681 (1985).

22. E. Mercado, III, R. Uyeyama, A. Pack, and L. Herman, Memory for action events in the bottlenosed dolphin. *Animal Cognition*, **2**, 17–25 (1999).

23. E. S. Savage-Rumbaugh, Language learnability in man, ape, and dolphin, in: *Language and communication: Comparative perspectives*, edited by H. Roitblat, L. M. Herman, and P. E. Nachtigall (Earlbaum, Hillsdule, NJ, 1993), pp. 457–484.

24. D. Reiss and B. McCowan, Spontaneous vocal mimicry and production by bottlenose dolphins. *Journal of Comparative Psychology*, **107**(3), 301–312 (1993).

CHAPTER FOURTEEN

Alex: One Smart Parrot

Irene Pepperberg has a Ph.D. in theoretical chemistry, which makes one wonder what possessed her to devote her professional life to teaching parrots. Here is her reply (personal communication, 2001).

> I had always been fascinated by avian intelligence and avian communication, ever since I was an only child, living above a store, with a budgie as my only playmate. The impetus to start working in the field, however, came when I saw the early Nova programs on the signing chimps, the work with dolphins, and the program "Why Do Birds Sing?" Having gone to MIT, where biology, at the time, concentrated on cellular and neurobiology, I was unaware that serious research was possible in the field of animal behavior. I was fascinated by the field, particularly with the story of Peter Marler, whose first doctorate was in chemistry, and who decided to study bird song after hearing chaffinch dialects as he collected soil samples for his degree. I was not very happy as a theoretical chemist, so I began to audit courses in bird behavior and child language, read incessantly in the libraries, and attended every seminar on behavior I could find in the greater Boston area. I'd spend 40 hrs/wk on my doctorate and another 40 hrs/wk learning behavior.

A VISIT TO THE PEPPERBERG LAB

And learn behavior she did, which launched her study of parrots. Beginning at Purdue University, then with a stint at Northwestern University, she moved to the University of Arizona. I visited her laboratory there, which was like a trip into another dimension. The strangeness started when I went down a ramp from Lowell Street and came out on the second floor of the biology building, when I expected to be in the basement. It continued when I knocked on the double door of room 235, and Nicole McReady, one of the lab assistants, opened it.

As a privileged guest, I was admitted to the inner sanctum of—what?—three little African gray parrots (*Psittacus erithacus*, to be fancy about it) standing around, Griffin and Kyaaro (Kyo for short) on a counter and Alex on a lower table in a very ordinary-looking room that was designed to be an introductory biology lab. Newspapers littered with discarded parrot food lay about, under or in the vicinity of each parrot perch. To the right along the north wall was the refrigerator used to store the parrots' food. At the northeast corner of the room is the door to the room where Alex sleeps, and standing on the east wall is a table with a cage where Monty, a parakeet, lives.

The renowned Alex has intellectually gone where no other parrot has gone before. My feeling of 4th-dimensional strangeness in this very ordinary room comes from my awareness that Alex can answer questions about the name, color, shape, and composition of some 100 objects, usually presented in groups of seven. He is, then, in my mind an intelligent being, not really a parrot, but an alien life form on this planet. Griffin and Kyo enjoy a guilt-by-association attribution of intelligence on my part, and I therefore feel myself being observed, examined, and evaluated by three alien intelligences. In quieter moments I look back on this scene and realize that it really differs little from being evaluated by three strange animals of any kind, dogs for example, but in this weird second-story basement with three animals of a species that I had never encountered before, I was acutely aware that humans were not the only intelligences in the room.

Dr. Pepperberg, absent when I arrived, soon returned, and proved to be a friendly and accepting hostess. However, I had come at a bad time for her; she had just heard that her application for support from the National Science Foundation had been denied. I hoped that she would not associate me with that rejection, and follow the hallowed "kill the messenger" tradition. However, a minute later she was opening mail and found a check for $1,500 that a friend had sent; that brought tears, and she said she would manage to keep her work going somehow.

Dr. Pepperberg is not an overprotective parent to her parrots; she encouraged me to interact with them if they were willing. Nicole told me that Kyo was unlikely to be friendly, and might bite if pressed, and that Griffin was also likely to be standoffish. Alex, on the other hand, already was moving on his table to get closer to me. I was uncertain about how to respond. At that moment Dr. Pepperberg's veterinarian friend, Dr. Rand, arrived and Alex climbed aboard his hand and transferred to his shoulder. Dr. Rand explained that

Alex was fond of him, and of most human males, and liked to court him. An uncomfortable part of Alex's courtship is that gray parrots regurgitate food as part of their courtship ritual, and Dr. Rand was afraid that he might be so honored if he allowed Alex to court him for very long. Because of his preference for courting males over females, the lab joke is that Alex is a gay (in addition to gray) parrot. Maybe Alex dislikes the smell or taste of the perfumes or body lotions that human females are more likely than males to use, or maybe he had been hand raised by a male prior to the time Dr. Pepperberg acquired him.

After Dr. Rand left, I pulled up a chair beside Alex's table and sure enough, he came over to the hand that I offered. I thought he was about to climb up, but what he wanted was to engage in more courtship. This he did by gently fondling my fingers with his beak and tongue, and treading on my lower fingers with one foot (which I misinterpreted as ambivalence about being picked up, but learned that it is really part of the courtship ritual). Dr. Pepperberg has acquired only males just by chance. Whether the presence of a female would lead to conflict between the males is unknown, but from what I saw I would guess that it would be easy to get the parrots to breed!

Alex Presents a Conflict

When everyone, including the three small aliens, had settled down, Alex launched into repeated requests: "Want corn." "Grain." "Grapes." "Want cork." Over and over and over he repeated the requests. When Nicole and Irene were not otherwise occupied, they treated each request as real, and offered the requested item. I soon got into the act as well, and after all initial requests had been honored, when Alex asked for items that he already had, we pointed out to Alex that each item was already on his table. Repeated offering of the items nearly always led to rejection. Irene told me that Alex really liked fresh corn, and only the cooked version was available in the lab, so that was a possible reason that he was rejecting the corn. However, I soon concluded that Alex did not really care about any of the items he was asking for; rather he was controlling his audience, gaining attention and enjoying the spectacle of three humans dashing about, opening and closing the refrigerator and trying to grant his every wish.

That is the conflict presented by Alex. Unless animals' vocal or signed requests are treated as genuine, they will not be motivated to learn. Under that regimen Alex learned to say "want corn" in order to get corn, "want grapes" to get grapes, etc. But giving him the reward could not be accomplished without certain side effects. Alex soon learned to generalize these responses from the training situation to the everyday life of the laboratory. In that situation the first consequence of each request is that Alex makes his minions move about and attend to him. Thus he has learned that vocalizations whose ostensible meaning is that he wants various objects are useful for obtaining non-object-related attention. And Alex appears to be a social animal, liking interaction with humans better, so far as I could tell, than interaction with other male parrots. The same conflict must be a consequence of the same training philosophy with other social animals—for example, with human children, who are notorious for using crying to gain attention, as well as for obtaining other desiderata, milk and clean diapers, for example. Watching Alex is instructive because we can be a little more objective about his responses, and a little more certain that he is using his vocalizations for a purpose that we did not intend, than we can be with children, especially our own children! Too, because Alex repeats this routine with every novel visitor, he may also be trying to learn whether a stranger in the lab shares his repertoire.

Afternoon with Alex

After lunch Nicole, the single dedicated caretaker during the morning, was replaced by Robert Sandefer, another undergraduate student at the University of Arizona. The laboratory was unusually quiet because it was the between-semester Christmas vacation, and most students had left Tucson to be with their families. During the semester there would have been multiple daily projects teaching or testing Alex, Griffin, and Kyo. Each project requires the participation of at least two humans, and usually two projects are under way simultaneously. Thus about six people, plus the three birds, would usually be present. I was very fortunate to be there at a quiet time, when my presence was probably desirable to keep the birds from being too bored.

Boredom is a problem for all laboratory animals. In chimpanzees it leads to rocking, hair plucking, and often to extreme withdrawal and low reactivity. In parrots boredom may be a cause of feather plucking. Alex, who was 23 years old, still retained many of his feathers; he is probably most acclimated to laboratory life, and, because of his more extensive repertoire and greater extroversion, he probably gets the most attention. Kyo, 9 years old, had even more feathers than Alex. Griffin, $4\frac{1}{2}$ years old, had plucked out so many of his feathers that he looked more like a young owl than a parrot; he was covered only with small downy feathers over most of his body. All three parrots are nearly the same size, no more than 9 or 10 inches high in their normal upright posture. Alex would never allow me to pick him up, so I could only guess at his weight; my guess is less than a pound! Dr. Pepperberg says that Griffin weighs 500 grams, very close to a pound.) Thus my guess could be wrong, but not grossly wrong, and Alex's brain cannot be much larger than a shelled walnut. How can an animal with a brain so small be so apparently intelligent? How can he answer questions about the materials of which objects are made (metal, paper, wool, or rock), or colors, or shapes? He must first understand the question, so that he does not tell you about color when the question was about material, or vice versa, or about shape when you wanted to know about color.

Some critics have alleged that "Pepperberg is cuing her birds so that they give the right answer." The allegations seem unlikely to be correct; the cuing would have to be nearly as complex as understanding the question and knowing the answer, and therefore nearly as interesting to study. While pondering that issue, I tore off a small chunk of the newspaper that was on Alex's table to catch discarded food, folded it to make it a bit thicker, and, using the procedure that I had seen applied to Alex, rubbed it against his beak while asking "What matter?" (Alex understands "matter" to mean "material.") Alex unhesitatingly replied "paper." If there are ways to cue Alex, I do not know them. To be cued, he would have to have read my mind. It's more likely that he identified the category that I was asking about (material, not color, or shape, or the name of an individual toy, which he also knows), and then picked the correct answer, based on the exemplar that I used (a torn-off bit of newspaper), which was an entirely new example of the paper category. (I am entirely convinced that Dr. Pepperberg and her students are not cuing Alex. Any claim that they are cued is not only false but also ridiculous, especially when one considers the elaborate precautions Pepperberg has taken against cuing.) Her book, *The Alex Studies*[1] describes the precautions in detail; I summarize them below.

It does seem impossible that Alex knows what he knows, but he knows it anyway. It is impossible that an animal so stupid that he will court a human hand for hours can also answer questions from several categories about over 100 items, but it's true, anyway. We are apparently deep in the 4th dimension here, and amazing things can happen if you have extra dimensions. These little aliens are a paradox. With their tiny brains, they could not possibly do what they do. But they do it anyway. I wish I could be around when we finally figure out how small brains accomplish so much, but I do not think I will be. Maybe my astonishment is more fun than my understanding would be. We humans with our huge brains are the stupid—or at least inefficient—ones. Compared to Alex, we are indeed brilliant; but we should be a lot more brilliant than we are! The daily headlines tell us that planes are hijacked, hostages killed, and huge buildings brought down, that bombs are thrown at little Catholic girls on the way to school, that yet another Palestinian suicide bomber has died along with several Jewish victims, etc. if Alex understood this, he would say, with parrot brevity, "What sense?"

The following morning a photo shoot was scheduled. A photographer was to take pictures to accompany an article in *New Scientist*, a British journal. When I arrived the shoot was well underway. I think photographers have to be obnoxious in order to get their work done. At any rate, this one was, just like the last one I had encountered before him. Dr. Pepperberg was a patient and perfect model, following orders to "relax your face more," "turn a little to the left," "move over here and hold the parrot lower," etc. The photographer's assistant was equally compliant, handing him equipment and taking exposed film at a high rate. The parrots were quite another matter. They did everything that Dr. Pepperberg wanted them to do, but accompanied it with a barrage of querulous "Wanna go backs" in an attempt to convince those present that they had had enough of this nonsense. She continuously reassured them that it would not be much longer before they could go back—but of course it was quite a bit longer. Finally the ordeal was over, and they were allowed to go back to their perches. The photographer was delighted with the results, and no doubt *New Scientist* was equally pleased. The parrots were not pleased. I suppose posing is the price of celebrity. Later, when I saw an article based on Alex's accomplishments (not in the *New Scientist*) with the ridiculous title "Polly wanna' Ph.D." I wondered whether the price of celebrity was too high.

Soon after the photo shoot, it was time for me to leave. Back to a world with only three dimensions.

ALEX'S ACCOMPLISHMENTS

Alex performs intellectual feats that, until he demonstrated them, were regarded as impossible for such a small-brained animal. Pepperberg[2] describes one of the tasks:

Alex was to be presented with pairs of objects that could differ with respect to three categories: color, shape, and material (e.g., a yellow rawhide pentagon and a gray wooden pentagon, a blue wooden square and a blue

Dr. Pepperberg asks Alex a question about a typical tray with seven items varying in color, shape, and material. Alex must identify which of three question types was asked, about what object, and then find the correct answer among the alternatives available to him. Photo by William Munoz.

paper square). He would then be queried "What's same?" or "What's different?" The correct response would be the label of the appropriate *category*, not the specific color, shape, or material marker that represented the correct response (e.g., "color," not "yellow"). To be correct, therefore, Alex would have to

1. attend to multiple aspects of two different objects,
2. determine, from a vocal question, whether the response was to be based on sameness or difference,
3. determine, based on the exemplars, what was the same or what was different (e.g., were they both blue, or square, or made of wood?), and
4. produce, vocally, the label for this particular category (pp. 491–492).

Alex answered these questions with about 75 percent accuracy. This is amazing, not only given what we thought we knew about the limitations of avian intelligence, but also given that David Premack, who devoted considerable time to teaching the concepts of same and different to chimpanzees, thought it likely that only primates could acquire these concepts. Furthermore, Premack believed that learning the concepts of same and different produced the only cognitive effect of language training.

Alex had already[3] provided evidence that he had mastered the class concept; that is, he could label the colors, shapes, or materials of which objects were made, upon being asked "What color?" or "What shape?" or "What material?" Answering these questions correctly about novel objects requires that the respondent, whether human, chimpanzee, or parrot, choose from the set of answers that correspond to the question asked. If the questions were asked about familiar objects, it would be possible (although perhaps unlikely) that the respondent had simply learned to make the correct response to each question in the presence of each object. This stimulus-response account cannot, however, be applied if the respondent has never before responded in the presence of the object. Alex later achieved similar accuracy when required to search a set of seven objects in order to answer more complex questions like "What is the color of the wood?" His overall accuracy was over 80 percent on this task, and over 75 percent the first time he was asked such questions in the presence of a new array.

Alex is familiar with over 100 different objects, and the typical set presented to him consisted of seven objects on a tray covered by a green cloth. Alex could name seven materials: hide, wood, rock (playdo), cork, paper, chalk, and wool; seven colors: green, rose (red), blue, yellow, gray, purple, and orange; and five shapes: 2-corner (football-shaped), 3-corner (triangular), 4-corner (square), 5-corner (pentagonal), and 6-corner (hexagonal). It is an interesting exercise to calculate the probability that Alex would randomly generate the correct response to a question like the above, "What is the color of the wood?"

First Alex had to select a category from which he would choose the appropriate response. Four possible categories of questions were used, corresponding to two of the three categories above (color and shape), plus two additional kinds of questions, one asking him to name the object in the set of seven that had a particular color, and the second asking him to name the object with a particular shape. Therefore, it is reasonable to assign a value of $\frac{1}{4}$ to the probability that Alex would select the correct category by chance.

Assume that Alex has been asked the color of an object. Then the probability that he would, by chance alone, select the correct object to analyze is $\frac{1}{7}$, given that there were seven objects on the tray. Finally, to select the correct color from the seven color names in his repertoire, he would again have one chance in seven of doing so by guessing. We could therefore calculate the overall probability of a correct answer, by chance, as being $\frac{1}{4} \times \frac{1}{7} \times \frac{1}{7} = \frac{1}{196}$. That is, indeed, the correct value if Alex were forced to approach the problem in this stepwise manner.

However, Alex need not have paid any attention to the objects on the tray before making a response. If he got through the first step, selecting a category, then he has a $\frac{1}{7}$ chance of selecting the correct response without identifying the object to which the question refers! Thus a more reasonable chance probability is calculated by completely ignoring the intermediate step, for a chance probability of $\frac{1}{4} \times \frac{1}{7} = \frac{1}{28}$. Another way of proceeding would be to ignore the question altogether and simply generate a random response from his vocabulary, which contains over 80 words; the probability of being correct in this case would be less than $\frac{1}{80}$.

If Alex were to be correct as often as he actually was, however, he could not have ignored any step. He *had* to identify the category and the object in order to answer the question about its color, or he had to find the object with the asked-for color (or shape, depending on the question) in order to answer questions at the high level actually observed in the study.

It is interesting to ask what is the best that can be done under differing assumptions about the abilities of the subject (or the abstract model representing the subject). Assume that the subject cannot identify the response category (color, shape, object-with-color, object-with-shape) correctly, but can find the right object and select the right response from the randomly chosen category. Then clearly the probability of being correct is $\frac{1}{4}$, provided that equal numbers of each question type are asked and that random choices of a category are made.

Suppose instead the subject can pick the category correctly and can read off all the properties of all the objects on the tray perfectly, but has not a clue about identifying the correct object on the tray. Then, given that all objects on the tray differ in the attribute being asked for, the probability of being correct, by chance, will be $\frac{1}{7}$. The actual situation for Alex was slightly more complex; for one thing, he had only five names for shapes, so if the question were about shape and Alex correctly selected that category, his chance probability of being correct became $\frac{1}{5}$.

Finally, if Alex could get through the first two performances correctly—that is, could identify the response category and the object asked about, but had not a clue about how to select the correct response from the category, the probability of being correct is again $\frac{1}{7}$ or $\frac{1}{5}$, depending on the question. We must, therefore, conclude from his actual level of performance, above 80 percent correct, that he could not have been seriously deficient in any aspect of the performance. He was able to understand what category of answer was required (color, shape, object-with-color, object-with-shape), to select the object on the tray that the question was about (2-corner hide, or yellow wood, or any of the seven objects on the tray), and finally to select the name of the attribute (for example, rose, if the color of the object was the attribute requested). Pepperberg[4] reports that the probability that Alex's performances on questions of this type that he had never been asked before would have occurred by chance was less than 1 in 10,000, statistically very unlikely indeed!

[Alex has a vocabulary of more than 80 vocalizations, although some of them are approximations of their English equivalents; for example, Alex says "Mah-mah," rather than "material." That is taken as a correct response to "What's same," if he is responding to a pair comprised of a green wooden triangle and a blue wooden square, or to other pairs made of the same materials.]

MAKING SURE ALEX IS NOT RESPONDING LIKE CLEVER HANS

Pepperberg, like all other contemporary language researchers, is acutely aware that she must not be fooled by the Clever Hans effect. She takes great pains to avoid that. When Alex is being tested, she is present, but has her back turned to Alex.[1] She does not know the correct answer to the question, so she can not be biased toward hearing what she expects. She vocally repeats the answer that she hears Alex produce; the person in Alex's view then judges Pepperberg's response as matching or not matching the correct answer and rewards, or does not reward, the parrot accordingly. The person (usually a student) administering the test has not been involved in training the skill being tested. Finally, it is far more difficult to imagine how cuing could direct a choice among available responses than to imagine how it could stop a "go-no go" response like the tapping response that was used by Clever Hans.

Pepperberg's goal is to leave no choice but to believe that Alex understands the questions asked of him and answers on the basis of that understanding. I think that she has succeeded, and that Alex has demonstrated that the apparently impossible was, after all, possible. Thus Pepperberg is to be congratulated for the spectacular success of her training program with Alex. Let us examine the broad outlines of the program that brought about this success.

ALEX'S TRAINING

To start at the beginning, Alex was purchased from a Chicago-area pet store in June, 1977. He was then about 13 months old and had had no formal vocal training. Pepperberg is blessed in one respect; Alex is much easier to maintain than most animal students of language. Here is what she has to say about Alex's care:[2]

> To give this subject some control over his environment while maintaining experimental rigor, my students and I allow him free access, based on his vocal requests, to numerous areas in the laboratory room while we are present (~8 hours/day); in our absence he is confined to a cage (~62 × 62 × 73 cm) and the desk on which it rests. Water and a standard psittacine seed mix (sunflower seeds, dried corn, kibble, oats, etc. [this mix has been replaced by a pelleted diet]) are continuously available throughout the day; fresh fruits, vegetables, specialty nuts (cashews, pecans, almonds, walnuts) and toys are used in training and are provided at the bird's vocal requests. (p. 476)

TRAINING PARROTS VERSUS TRAINING CHIMPS

Although caring for Alex is probably no picnic, neither is it comparable to the rigors of caring for a great ape. Alex is no general threat either to his caretakers, who need have little concern about losing fingers, nor to his surroundings, although Alex is no doubt capable of picking apart both fingers and furniture. Nevertheless, his accommodations need not be as sturdy as those that must be constructed to contain great apes. Given all of this, Pepperberg made a wise choice of subject!

Pepperberg's choice of a general training regimen was also very astute. She called her procedure "The model/rival approach, using intrinsic rewards," and describes it as follows:[2]

> Our training program emphasizes the use of live, interacting tutors and intrinsic, rather than extrinsic, rewards (Pepperberg, 1981). Many previous programs designed to develop communicatory skills in both human and nonhuman subjects relied on noninteractive techniques and rewards that were neither directly related to the skill being taught nor varied with respect to the task being targeted. Our procedures focus instead on demonstrating referential, contextual use of each targeted vocalization and on using as exemplars those objects and actions that themselves arouse the interest of the subject ... (p. 476).

THE MODEL/RIVAL APPROACH

The primary technique, called the *model/rival* or M/R approach, employs:[2]

> ... humans to demonstrate to the parrot the types of targeted interactive responses. This procedure is based on a protocol developed by Todt (1975) for examining vocal learning in Grey parrots .... In the presence of the bird, one human acts as a trainer, and a second human acts as a trainee. The trainer presents objects, asks questions about these objects, gives praise and reward for correct answers, and shows disapproval for incorrect answers (errors similar to those being made by the bird at the time, e.g., "Wood" for "Green wood"). The trainee is both a *model* for the bird's responses and a *rival* for the trainer's attention. The roles of model/rival and trainer are frequently reversed to demonstrate the interactive nature of the system, and the parrot is given the opportunity to participate in these vocal exchanges. (p. 476)

Pepperberg's use of the model/rival approach somewhat parallels the situation in which Kanzi learned while Dr. Savage-Rumbaugh was trying to teach Matata. As we noted earlier, Kanzi as an infant was present as experimenters tried to teach Matata how to use lexigrams. Matata was an extremely poor student, and probably did not provide a very effective "rival"; she was certainly an almost complete failure as a "model." In one respect, however, perhaps she was unusually effective; she was given over 30,000 trials in the attempt to teach her about lexigrams, which gave Kanzi ample opportunity to observe the connection between pressing a lexigram and the consequences that ensued. In any case, Matata and Sue Savage-Rumbaugh comprised a social unit with two entities trying to exchange information, just as in the Todt procedure.

After observing Matata in the learning situation for about 2 years, Kanzi's own language training began. From that point on, he progressed very rapidly in the acquisition of lexigrams, using an informal regimen that emphasized natural interaction with experimenters in contexts selected to be interesting to Kanzi. Although this approach was neither modeled on, nor similar in detail to, the model/rival approach, it shares with it the emphasis on observational learning and social interaction involving situations of interest to the learner.

The model/rival approach contains many of the same elements that are present in the environment of human children in two-parent families, or in exchanges between one parent and an older sibling, close friend, or neighbor. The humans engage in frequent linguistic interactions in the presence of the child, and, as in the model/rival technique, they often change roles. The child is usually free to join the conversation, as Alex was. Intrinsic reinforcers are often given to human children, although in a less formal way than they were given to Alex. Of course, the usual human interaction is far less formalized than the model/rival approach conceived by Todt and used in modified form by Pepperberg; it may, for that reason, not be as effective as Pepperberg's version.

Pepperberg used an interesting procedure for reinforcing Alex's correct responses. When Alex vocalized the name of an object, he was given the object. When objects were novel, or when Alex was interested in the object for some other reason, this worked quite well. However, if the object was very familiar, or was a food of which Alex was tired, he had little or no interest in it. Repeated no-interest trials could lead to boredom or lack of cooperation. An example of this problem was the attempt to teach Alex the word bread.

After a time, Alex had no further desire to eat bread, and his rendition of the word never became perfect.

To circumvent this type of problem, Alex was allowed to request another object after giving the correct answer to a question concerning the non-preferred object. For example, if Alex were shown bread, asked "What's this," and responded "graed," one of his approximations for "bread," he could then ask for green beans, a type of food that he liked better, and receive green beans.

Pepperberg[5,6] has begun to determine what the necessary and sufficient conditions are for the model/rival technique or related techniques to work. She believes that three features of a teaching situation, derived from Bandura's social modelling theory, are critical for the learning of a human-derived code by parrots and non-human primates. First, the teaching must be referential; that is, the symbol and the object or situation must be presented together so that the learner has a chance to see their relationship. The less ambiguous the referential relationship, the better; for example, it works better if the object for which the symbol stands can be presented as a reward for correctly vocalizing the name of the object than if some single reward is used for all correct vocalizations. In the latter case, the learner could suppose that the symbol stands for some aspect of the reward, rather than for the object referenced. This is reminiscent of the early difficulties in trying to teach Sherman and Austin to name objects by rewarding them with a single reinforcer.

Second, the vocalization should be functional. In a simple situation, the vocalization could enable the learner to obtain the object, so the reason for learning the sounds is clear. The trainer would "hand over" the object referenced upon hearing the correct vocalization, but not if the vocalization were absent or incorrect. During model/rival training, one of the tutors would hand the object to the other when a request was correctly made, and the learner can observe all the characteristics of such interactions.

Third, social interaction seems to be extremely critical. Even when reference was present with minimal functionality and minimal social interaction, there was no observable learning. The exact quantity and quality of social interaction that are required are not yet known, but some social interaction is clearly required, and it appears that more intense interaction is preferable to more placid interaction, at least up to a point. Apparently the social interaction need not be between tutors of the same species as the learner, because Alex learned from human tutors.

A juvenile parrot given audiotaped input that was neither referential nor interactive learned nothing about the English labels for objects on which all three parrots in the study were being trained. A second juvenile, even in a condition that provided referential information and some information about context, also failed to learn. A third, adult parrot given the usual interactive tutoring with limited context and reference by live tutors, learned to produce the labels, but showed no evidence of understanding them. Various other conditions, including videotaped presentations that presented some referential and functional information, failed to produce any observable learning. Thus the results to date indicate that fully representational, functional training in a situation involving substantial social interaction is needed to produce correct vocalizations with understanding. The full model/rival approach with complete access to the objects being referenced and to the context in which the communication occurs was necessary for best results.

Like human children, Alex sometimes practiced by verbalizing when he was by himself.[7] The words that he was being taught at any particular time comprised only a small percentage of his solitary speech, but sounds related to those being taught did tend to occur at higher frequency when he was learning the related sound than when he was not. Alex was more active during daylight hours, so his speech was recorded during early morning and early evening hours, before his trainers arrived for the day or after they left. These authors suggest that the observed playing with sounds may occur because it allows children and parrots alike to practice without receiving negative feedback for errors. Children have been observed to engage in such solitary word play through age 7; Alex was 10 when Pepperberg et al. conducted their first study of his solitary verbalizations, and 12 when the second set of observations was collected. It thus appears that Alex might be a useful model for trying to determine the conditions under which children engage in word play and for trying to determine whether such word play is helpful in learning new words.

Pepperberg and Sherman[8] recently applied their model-rival approach to teaching children who were autistic, developmentally delayed, or hyperactive. The results were quite encouraging; all of the participants improved both their social and communication skills and their use of contextually appropriate behavior. Thus, at least for these participants, bypassing the vocal channel, as the other animal researchers have done in teaching children, was not necessary.

✍ SUMMARY

A summary of Alex's accomplishments[1] includes all of the following

1. Alex could identify, refuse, and apparently request objects for play or for food; he could generalize label use for objects that differed from training objects.

2. He could select his responses from the correct category (shape or color), depending on the question asked.

3. He learned the concepts of same and different.

4. He learned to respond to the absence of information (for example, to say "none" as the answer to "What's same?" when two objects differed on all dimensions).

5. Alex can label the number of up to six objects or subsets of objects in an array. For example, Alex was shown an array consisting of one piece of orange chalk, two pieces of orange wood, four pieces of purple wood, and five pieces of purple chalk. He was then asked "How many purple wood?" He answered correctly, "Four."

6. Alex can label an attribute of an object defined by another attribute; for example, he can find the object that is made of wool and provide its shape in order to answer the question "What shape is wool?" Alex extended this ability to situations demanding that he consider two attributes in order to identify the correct object, as in answering the question "What color is the three-corner wood?"

7. Alex can state which of two novel objects is larger or smaller (or indicate that they are the same size) in response to questions like "Which color bigger?" or "Which matter smaller?"

8. Alex passed advanced tests of his concept of object permanence. In one such test an object is obscured by the experimenter's hand or a cup and passed under one or more screens. Then the subject (Alex, in this case) is shown that the hand or cup is empty, and must immediately remove the correct screen in order to obtain the hidden object. In a final twist on this test, a less desirable object was substituted for the object originally shown (a pellet rather than a cashew nut), and Alex indicated anger at the substitution after selecting the correct screen.

9. Alex learned the difference between merely naming an object and requesting an object that was not present by preceding the name with "want," as in "want corn." He also generalized it to other situations, as in "Wanna go

back." Remember Alex's reaction to the photographer during the visit to the laboratory described early in the present chapter?

A fair summary of Alex's accomplishments is that he learned the functional use of a vocabulary of over 80 words, and his ability is comparable to that shown by the great apes—an amazing demonstration of unsuspected cognitive ability in a bird! Alex's use of "want" followed by the object or activity desired demonstrates a rudimentary syntax.

POSTSCRIPT

In mid-2000 Dr. Pepperberg moved Alex and Griffin to the Massachusetts Institute of Technology. That is home for Pepperberg. Presumably her laboratory is not on the second story, underground, as it was in Arizona, but the presence of parrot language research in a school known for engineering and hard science is still other-worldly. Just the place for little aliens.

REFERENCES

1. I. Pepperberg, *The Alex studies* (Harvard University Press, Cambridge, MA, 1999).
2. Irene Pepperberg, Conceptual abilities of some nonprimate species, with an emphasis on an African Grey parrot, in: *"Language" and intelligence in monkeys and apes*, edited by S. T. Parker and K. R. Gibson (Cambridge University Press, NY, 1990), pp. 469–507.
3. Irene Pepperberg, Cognition in the African Grey parrot: Preliminary evidence for auditory/vocal comprehension of the class concept. *Animal Learning & Behavior*, 11, 179–185 (1983).
4. Irene Pepperberg, Cognition in an African Gray Parrot (*Psittacus erithacus*): Further evidence for comprehension of categories and labels. *Journal of Comparative Psychology*, **104**(1), 41–52 (1990).
5. I. M. Pepperberg, Vocal learning in grey parrots (*Psittacus erithacus*): Effects of social interaction, reference, and context. *Auk.* **111**(2), 300–313 (1994).
6. I. Pepperberg, Social influences on the acquisition of human-based codes in parrots and nonhuman primates, in: *Social influences on vocal development*, edited by C. T. Snowdon and M. Hausberger (Cambridge: Cambridge University Press, Cambridge, 1997).

7. I. Pepperberg, K. J. Brese, and B. J. Harris, Solitary sound play during acquisition of English vocalizations by an African Grey parrot (*Psittacus erithacus*): Possible parallels with children's monologue speech. *Applied Psycholinguistics*, **12**, 151–178 (1991).

8. I. Pepperberg and D. Sherman, Proposed use of two-part interactive modeling as a means to increase functional skills in children with a variety of disabilities. *Teaching and learning in medicine*, **12**(4), 213–220 (2000).

CHAPTER FIFTEEN

Evaluations of the Ape Language Research

Some criticisms have been applied to most, if not all, of the ape language research. Others apply only to one of the methods of circumventing the vocal channel or to a specific research project.

In the first category, critics have alleged that the apparent comprehension shown by all language-trained apes is a result of the Clever Hans effect. However, all of the ape language researchers have been so acutely aware of this possibility, and have been so careful to control for it, that this criticism need no longer be taken seriously as a blanket criticism of the research. In saying this, I am not denying that apes can take advantage of nonverbal cues; indeed, they are masters of that art.

What I do claim is that: first, ape language researchers have demonstrated that ape performances cannot be attributed solely to nonverbal cues. Second, the extent of nonverbal cuing that would be needed to guide the observed performances would have to go far beyond the cues needed to start and stop a response; that is, the cues would have to guide the *selection and execution* of one out of many responses, not just start and stop a response. (Contrast Clever Hans' need to stop after, say, 42 hoof taps with the information Kanzi needed to carry out the command "Put the pine needles in the refrigerator," a response that he had never performed or seen anyone else perform.) Third, nonverbal cues are a part, a legitimate part, of all face-to-face human communications, and should not be routinely excluded from animal-animal or animal-human communication. Thus the Clever Hans phenomenon is probably more useful to remind us of that fact than it is as a criticism of language studies.

255

APE PERFORMANCES AND CLEVER HANS

Videotapes of Austin and Kanzi are very instructive in this connection. When Austin (who did not understand English, at least when he could not see the speaker) was asked to point to, or hand over, an apple or some alternative object, he was notably deficient in understanding, and the direction of his gaze was toward the experimenter, not toward the objects about which he was being asked. Austin was doing his dead level best to get cues from the experimenter—to be the chimpanzee equivalent of Clever Hans, Clever Austin. When the experimenter in view of Austin knew which object was requested, and Austin could see him or her articulating, Austin did rather well. When he had to rely solely on auditory information delivered via headphones, he failed miserably. This was perfectly clear to the experimenters, who thus made no claim that Austin understood English. Premack, among others, forcefully made the point that it would take a very stupid or very deluded experimenter not to notice when apes were seeking nonverbal cues.

In stark contrast, when Kanzi was asked to choose the lexigram that represented a word spoken to him in English (grape, for example), his attention turned immediately to the lexigram board, away from the experimenter who stood behind him. And Kanzi, unlike Austin, was impressive in his ability to choose the talking lexigram that represented the English word. Similarly, when the Gardner's chimpanzees were being tested on their ability to give the appropriate sign for objects displayed on a screen, their attention was unwaveringly on the screen. They made no attempt to get cues from the experimenter, who could not see the screen, but could see what the chimpanzee was signing.

It is easy in the abstract to maintain that language-trained animals are taking cues from the experimenter to direct their responses. In the case of Austin, he was indeed doing his best to look for such cues, sometimes successfully. In the case of Kanzi or the Gardners' chimpanzees, it is easy to see that they are not using such cues. It really is true of the Clever Hans effect that seeing is believing.

The evidence of understanding provided by Kanzi, Panbanisha, Koko, Washoe, Chantek, Moja, Akeakamai, Alex, and other animals is so overwhelming that even a hardened skeptic must be swayed if he or she examines the evidence at first hand.

DOUBTS ABOUT WHETHER SIGNS, ETC., FUNCTION AS WORDS

A second very general criticism is that there is insufficient evidence to conclude that animals understand signs, lexigrams, plastic symbols, or even English words to refer to objects, actions, or grammatical signs. This criticism is more difficult to counter, but we have seen several lines of evidence that some animals know that their symbols refer to something else, that there is a mapping between symbol and object or symbol and action. One of the most convincing demonstrations was when Sarah indicated that her symbol for apple was round, red, and had a stem, when the symbol had none of those properties. An equally convincing demonstration was Sherman and Austin's nearly 100 percent success in classifying symbols as food or tool when they had not been taught which category the test symbols belonged to. We can add to these two demonstrations many others that are less rigorous; for example, when animals use a symbol to request a specific food, they nearly always select that food when more than one food is offered, sometimes when a more preferred food is among the offerings. This point can be generalized to actions; when Kanzi pressed CHASE and APPLE when he was a very young bonobo, he typically grabbed an apple, put on a play face, and turned to run away. Animals *do* understand the reference relationship between symbols and objects or actions.

CRITICISMS OF THE SIGNING PROJECTS

In the following discussion, I take the Gardners' work as the prime example, but several of the criticisms could be applied to all the projects in which signing was the method of communication.

A fundamental criticism of any scientific work is that the data are wrong or unreliable. The Gardners tried to exercise extreme care, but Pinker,[1] for example, is harsh on this point:

> The one deaf native signer on the Washoe team later made these candid remarks: Every time the chimp made a sign, we were supposed to write it down in the log .... They were always complaining because my log didn't show enough signs. All the hearing people turned in logs with long lists of signs. They always saw more signs than I did ... I watched really carefully.

> The chimp's hands were moving constantly. Maybe I missed something, but I don't think so. I just wasn't seeing any signs. (pp. 337–338)

Pinker's comment certainly encourages examination of the reliability of the Gardners' data; without such an examination, we could not be sure that their conclusions were not affected by distortions in the interpretation of the chimpanzees' movements. Alternative possibilities are that the deaf signer was looking for perfect signing, and did not recognize the modified forms that the chimpanzees produced. Of even greater concern, Pinker ignored information that was presented in a book published 4 years before his. In the follow-up of the Washoe work,[2] the Gardners report:

> At all times in the second project, there were several human members of the family who were deaf themselves, or who were the offspring of deaf parents, and still others who had learned ASL and used it extensively with members of the deaf community. With the deaf participants it was sign-language-only all of the time, whether or not there were chimpanzees present. The native signers were the best models of ASL, for the human participants who were learning ASL as a second language as well as for the chimpanzees who were learning it as a first language. *The native signers were also better observers because it was easier for them to recognize babyish forms of ASL.* [Italics not in original.] Along with their own fluency they had a background of experience with human infants who were learning their first signs of ASL. (pp. 14–15)

Pinker's criticism, although it deserves consideration, was based on the report of a single observer of the early work. The report is less dubious for the Washoe project, but it is certainly not damning for the Gardners' research in general. Several studies indicated that the reliability with which signs were identified was satisfactory, as I noted in the chapter on the Gardners' work. I also noted the need for videotaping signs so that reliability could be increased, and the Gardners and their coworkers did systematically videotape a large number of interactions. Without videotaping, it is difficult to follow the fast-moving chimpanzees' signing. Without videotape, it must be nightmarishly stressful to try simultaneously to observe, record, and converse in sign language, particularly for non-native signers. With videotape and several native signers, as in the Gardners' work with Moja, Tatu, Pili, and Dar, it was demonstrated that there was sufficient reliability of observation to support the conclusions offered.

However, critics have attacked the Gardners' conclusions as least as much as they have attacked their data. Many linguists believe that, whatever the chimpanzees did, they did not manifest language. Again, Pinker:[1]

> To begin with, the apes did *not* "learn American Sign Language." This preposterous claim is based on the myth that ASL is a crude system of pantomimes and gestures rather than a full language with complex phonology, morphology, and syntax. In fact the apes had not learned *any* true ASL signs. (p. 337)

We should be as critical of the critics as the critics are critical of the investigators; in this spirit, we wonder how Pinker could know that not one chimp learned any "true" ASL sign! On the very next page, Pinker cites Pettito's conclusion "Pettito estimates that with more standard criteria the true vocabulary count would be closer to 25 than 125" (p. 338). Although it is not clear from the context what chimpanzees Pettito was talking about, it appears that Pinker is accepting a contradiction to his assertion that no chimp learned any sign. But he might respond that he said *true* sign, by which he might mean that a sign had to be embedded in the full grammar of ASL; so our imaginary conversation could go on and on. The Gardners were wise to say that their interest was in whether chimpanzees could learn to communicate information to them, rather than in whether the chimpanzees' behavior conformed to some definition of "animals that had language." Pinker is simply wrong when he implies that the Gardners accepted "the myth that ASL is a crude system of pantomimes and gestures rather than a full language with complex phonology, morphology, and syntax." The mystery is where he got that impression.

Another severe critic of the ape language studies is Joel Wallman. In building a case against the significance of ape language studies, he traces the history of the emergence of words from non-word articulations[3] (pointing, exchanging objects, coordinating attention via gaze, etc.). Wallman defines a linguistic symbol as a physical marker for some mental representation (a definition that does not appear to be laudably operational). He discusses the advantages and disadvantages of iconicity, agreeing that arbitrariness is not a necessary characteristic of words.[3] "The transition to symbolic, referential usage entails a process of decontextualization, a psychological decoupling of word and object" (p. 52). Wallman claims that many of the signs used by chimpanzees are just "natural gestures," which are presumably not decontextualized enough to constitute true words.

[Wallman, like Pinker, cites the deaf volunteer who said that the apes were not signing, that the observers were "reading in" random gestures as signs.] He also claims that the chimpanzees did not understand the cheremic structure (location, shape, and motion of the sign), that they only crudely used "whole" signs without recognizing their underlying structure. The Gardners claimed that the errors made by Washoe and their other apes followed these structural lines, but Wallman disputed the claim.

He also criticized the Gardners severely for lumping together different places of articulation (according to ASL rules) in their attempt to show that many chimp errors were examples of the same class of sign, although the sign was scored as an error. He also was critical because they were lax in their requirements for hand shapes while signing, and because they discarded errors on which the two observers disagreed, without what Wallman believed was a good logical justification. Although there may be a sound basis for arguing that this class of errors should have been kept in the analysis, keeping them in could not have invalidated the conclusions; it might even have strengthened them! He makes this deadly confession[3] "There is no obvious reason to assume that inclusion of all errors in the analysis would have produced results less supportive of the Gardners' contentions. Nevertheless, this omission of data, apart from any of the other complaints above, would seem to invalidate the analysis" (pp. 63–64; a non sequitur indeed!)

Wallman is justified in asking whether chimpanzees' 80 percent correct labeling of slides of things like trees with signs demonstrates that the chimpanzees were referring with their signs. He says[3] "Yet a skeptic might characterize the animals' performance on these tests as demonstrating only a reliable (80 percent) rate of responding to classes of equivalent stimuli with rotely paired associates" (p. 64). Although this is possible, the statement might also describe a lot of human language!

Patterson and Cohn's work with gorillas was criticized for many of the same reasons as the Gardners'. The most salient criticisms besides those above include claims, especially by Herbert Terrace, that the animals are primarily imitating their keepers' signs, that they do not initiate conversations, that they use signs only to make requests, that the animals do not understand the symbolic nature of signs, and that the animals cannot construct sentences. All of these criticisms have been applied to the gorilla work, as they have to the chimpanzee work with sign language. In addition, Drs. Patterson and Cohn have been accused of being too willing to assume that what their gorillas sign makes sense, and too lax about accepting a sign as learned. These additional criticisms may or may not be justified, but, in either case, the attitudes targeted were intentional. In *The education of Koko* Dr. Patterson says:[4]

> And so almost from the beginning, as I developed procedures to map Koko's progress in a controlled and testable way that would produce credible data, I also tried to think of ways to capture and analyze those extra, teasing things Koko would do spontaneously. From the beginning, then, the language project with Koko was two experiments: the first a tightly controlled attempt to gather the facts about Koko's use of language, and the second a case study of Koko's use of the language we taught her. (p. 72)

It is the second experiment that has led to the strongest criticism from those who believe that science advances only through the first type of experiment. Time will tell which approach is most productive in the long run; meanwhile one wonders whether the critics have ever spent any time observing what language-trained apes do. My experiences led me to believe that they, with rare exceptions, have not. Koko would quickly convince the most stubborn critic that she initiates conversations far more than she imitates them. Critics might complain that she's too bossy!

CRITICISMS OF THE LEXIGRAM PROJECTS

The Language ANAlogue (LANA) project, like the Gardners' project, was criticized by skeptics like Sebeok and Umiker-Sebeok,[5] Wallman,[3] and then later Terrace.[6] Also, like all of the animal language researchers, Rumbaugh and Savage-Rumbaugh raised questions about their own research.

The fundamental question about the LANA project is whether Lana was generating sentences or just making a series of learned responses in order to obtain rewards. Chimpanzees are known to be very capable of making a series of responses. Wallman criticizes the gloss of the start key as PLEASE, although he acknowledged that Rumbaugh did not claim that every element in Lana's strings had linguistic significance.

Wallman noted that Lana's use of the period key, interpreted as erase, might have been the end of a sentence that Lana misconstructed without knowing it—so error data were eliminated by the experimenters' interpretation. It could be that Lana simply had made an error and thought the construction was correct. Wallman admitted that this criticism did not apply to the partial constructions offered to Lana by the experimenters, which Lana erased more often if they were incorrect up to that point.

Thompson and Church[7] found that 91 percent of Lana's utterances were one of six stock sentences that could have been response strings that she learned in order to obtain reinforcers, but they do not argue that all of Lana's utterances could be accounted for in this way. In addition, Pate and Rumbaugh[8] analyzed Lana's later productions and showed that the stock sentence interpretation was inadequate because there were 69 different sentences in Lana's later productions.

Wallman contends that Lana could have generated sentences, not by following grammatical rules, but by learning to press keys in a certain color order. The fact that von Glasersfeld had used differently colored keys for different grammatical categories gave Lana the opportunity to give the appearance of generating grammatical sentences by selecting fixed orders of colors.

This does raise an interesting question about how color might have been used by Lana. The obvious answer is that Lana's search for a lexigram that she wanted was made easier by dividing the keys into sets according to color. Thus if Lana wanted to address Tim, and knew that Tim's key was purple, she needed to search only among the purple keys. The same logic also applies to finding an action key that she wanted, for example, MOVE.

Thus one question is whether Lana was more likely to have thought "I need purple green brown yellow orange," or whatever color series would produce an acceptable sentence, or whether she was more likely to have thought "I need Tim move behind room" and used the colors to find the desired lexigrams more quickly. This suggests another question, "What cognitive process do humans or animals use to select words in the order in which they appear?" The translation from cognitions to grammatical sentences is the translation that Chomsky asserts takes place between the deep structure of thought and the surface structure of grammatical utterances. I think that this process remains rather mysterious for humans and animals. My introspections on my own translations from thought to verbalization lead me to doubt that Lana used color as the mechanism for ordering her lexigrams, but it remains a possibility until it is disproven by experiments in which the relationship between lexigrams and colors is varied, as the relationship between lexigrams and locations was varied in the Lana experiments.

Wallman also questions all of the anecdotal material gathered from Lana, pointing out that the nice locutions (like "apple which-is orange" for an orange) were often preceded by other locutions not so nice, as "apple which-is green." Perhaps Lana was generating random sequences, and finally hit upon an impressive locution. However, Lana often confused orange and green, so the substitution of green for orange may not have been random after all. In summary, Wallman says[3] "The Lana project was an innovative experiment that was flawed in method and misguided in conception.... The claim that Lana appreciated grammatical principles is undermined by the color coding of keys and indeterminate significance of Lana's use of the period key" (p. 37). This conclusion was probably foreordained by Wallman's negative bias, which led him to ignore or downplay evidence that is inconsistent with his conclusion. His own conclusion is undermined by the fact that Lana so consistently erased ungrammatical constructions. He might better have argued that Lana's initial failure in the naming task indicated that she did not know what she was saying; however, Lana again undermined the negative case by succeeding at the naming task once she learned what was expected of her. The Lana project was a success that fed on itself, in that it led to even greater linguistic accomplishments by bonobos like Kanzi and Panbanisha.

The later work with lexigrams, mostly by Dr. Savage-Rumbaugh, has suffered the same slings and arrows of the critics as the other ape language researchers, despite her own skepticism and, as a consequence, extra caution in proposing and supporting conclusions. We will again take Wallman as our designated critic.

Wallman sets out to savage the work with Sherman and Austin, in whom Dr. Savage-Rumbaugh believed that she had instilled true naming (referential use) of symbols. Sherman and Austin each learned to request tools from the other, with which the requestor could obtain food to be shared. However, Wallman points out that Epstein, Lanza, and Skinner arranged a similar interchange between two pigeons, one of which pecked a key labeled WHICH COLOR. The other pigeon looked at a color and pecked a corresponding R, G, or B, (for red, green, or blue) key, whereupon the first pigeon pecked a key labeled THANK YOU, which delivered food to the respondent. Then the questioner could peck the corresponding R, B, or G key and get food for herself. This was presented as parallel to the Sherman-Austin interchange. Wallman and other operant psychologists, presumably including Skinner, did not think that the pigeon interchange involved language, although it fits Skinner's own definition, "a response that requires for its reinforcement the intervention of another organism!"

The pigeon and Sherman-Austin paradigms share some features, but there are critical differences. It is not clear that there was any intent on the part of the pigeons to communicate. They did not share food, as Austin and Sherman did (sharing supports the belief that they were communicating intentionally). There is no reason to believe that the colors or letters referred to anything for the pigeons, but the referential function is clear for Sherman and Austin, who had to fetch a tool after seeing the associated lexigram. Thus the pigeon analog has entertainment value, but does not appear to be close enough to constitute a serious counterexample.

Wallman questions the conclusion that Sherman and Austin were able to classify lexigrams as standing for food versus standing for tools (or, more conservatively, those standing for food and those standing for non-food). Many of the objects used in the test were familiar, and had been used in earlier phases of training. But about half were novel, so Wallman has to concede that the corresponding results were valid. However, he likes Epstein's suggestion that the apes placed items that, through conditioning, made them salivate, in one bin, and those that did not in another. That would account not only for the correct responses, but also for Sherman's classification of the sponge with the food objects. Fortunately for her point of view, Dr. Savage-Rumbaugh in another study had the chimpanzees distinguish between food, drink, tool, and location, which Wallman had to concede invalidated the account in terms of salivation-non-salivation. One might also ask why apes would choose salivation versus non-salivation as the basis for classification, rather than obvious physical dimensions like colorful versus dull, or having sharp versus curved angles, etc.

In discussing Kanzi and Mulika, Wallman again makes much of the fact that most utterances are requests for something, which he says is not typical of child language. However, Kanzi often took the portable keyboard to a place distant from his human companions and "talked to himself." Wallman agrees that such usage is not instrumental, but he further argues that when Kanzi indicates where he wants to go, it is comparable to the behavior of a dog bringing a leash. In supporting this claim, Wallman says[3] "… it seems plausible that a dog might learn to enact different behaviors according to where it wishes to go" (p. 76). This appears to be just the kind of anecdote that Wallman would jump on if someone else suggested it, and it has the additional feature that it is only an imaginary anecdote! Despite his skepticism, Wallman is prone to conclude that Kanzi's behavior exhibits some features of language because he was not trained to use lexigrams through instrumental procedures, did not have to be tutored to press the lexigram corresponding to an object, and frequently uses lexigrams non-instrumentally. Finally, it is not obvious that an instrumental purpose makes the symbol non-linguistic; Skinner's definition of language, in fact, made instrumentality the very core of language which

"...depends for its reinforcement on the intervention of another organism!"

CRITICISMS OF THE PREMACKS' RESEARCH

Wallman is as critical of the work with Sarah as he is of all other animal language research (I get the impression that this is not a coincidence). He begins by questioning the semanticity of Sarah's plastic chips; for example, he points out that a typical sentence like "Mary give Sarah apple" conveyed almost no information that was not already provided by the context. The only giver present was Mary, the only receiver was Sarah, the only operation in many sessions was giving, and apple was one among a very few gift items. Furthermore, this was the only sentence construction being trained at the time, so Sarah needed only to complete the sentence by substituting the apple chip into the blank space. Wallman claims that this type of "gratuitous attribution" applied across all, or nearly all, of the constructions.

The next criticism is of the transfer tests; it would not be very impressive if Sarah only learned to fill in blanks in a situation in which she was well trained. Thus she was tested in new contexts to see whether she could transfer what she had learned. The data provided by Premack indicated very good transfer performance, but Wallman pointed out a potentially fatal flaw in the procedure: Premack gave Sarah a series of transfer trials, so she could have learned quickly what to do in the new situation, rather than transferring a skill from the old situation. The only true test of transfer is the first trial in a new situation. After that, correct performance could be a result of feedback from the first trial, and even monkeys can exhibit one-trial learning after sufficient prior experience with similar problems.

Wallman also denigrates Sarah's ability to respond to longer sentences like "Sarah cracker red dish candy yellow pail insert" because the Sarah chip is obvious and because she had been trained in a long sequence of trials, first on the component operations, then with both in sequence, and finally on the combined form. Thus she was first trained on "Sarah cracker red dish insert," etc.

Sarah was also trained on "conditionals" of the form "If Sarah take banana then Mary no give Sarah chocolate." Again much of this sentence is unnecessary. In other cases Sarah responded correctly to "If Debby give apple Mary then Sarah insert cracker dish," but in fact Debby always gave the apple to Mary, so Sarah only had to attend to the last, relatively simple, part of the sentence.

Wallman concludes[3] "Sarah's evanescent *oeuvre* entailed, at least, the development of skill in solving a number of problems devoid of linguistic significance and, at most, acquisition of a number of unrelated atoms of linguistic ability" (p. 45). Although Wallman's criticisms are, for the most part, reasonable, one of his final comments strikes me as a classic non sequitur. He argues that Sarah's and Lana's accomplishments can be downgraded because the systems they used (plastic chips and lexigrams) were learned by some language-handicapped children! He seems to assume that if a child could not learn language in the normal way, then anything he or she learned could not be language; hence what Sarah and Lana did could not be language either. The same argument could be applied to deaf children who use only American Sign Language! This is a depressing, as well as misguided, conclusion, denying that any new technique could help some human children to learn language.

A final irony, I hope, summarizes where we now stand with respect to all studies of animal language. In 1999 Dr. Sue Savage-Rumbaugh and Thomas Sebeok attended the Symposium on Animal Consciousness in Colorado. Dr. Sebeok had been a virulent critic of animal language studies and, with his wife, Dr. Umiker-Sebeok, had edited the critical book of 1980 that became infamous because of its condemnation of animal language research. I pick up the story from that point as Bill Fields tells it (Personal communication, Bill Fields, 2002):

> It was Sebeok and he was moving toward me and Sue... Sebeok kind of knelt down beside Sue. The non-verbal behavior was clearly warm and intimate. As I moved in to monitor their conversation, they both stood and put their arms around each other... I heard Sebeok so sweetly say, "Sue, I was wrong and you were right. You have endured and lasted despite all of the attacks on your work. You are a visionary and your work deserves recognition. I'm so sorry I didn't recognize this sooner."

I hope that the reconciliation of Sebeok, who died 2 years later, and Dr. Savage-Rumbaugh bodes well for a new era in animal language research, an era in which criticism is constructive and friendly, and acrimony is a thing of the past. And I hope the Israelis and Palestinians become fast friends.

REFERENCES

1. Steven, Pinker, *The language instinct* (Morrow, NY, 1993).
2. R. A. Gardner, B. T. Gardner, and T. E. Van Cantfort (Eds.), *Teaching sign language to chimpanzees* (State University of New York Press, Albany, NY, 1989).
3. J. Wallman, *Aping language* (Cambridge University Press, NY, 1992).
4. F. G. Patterson and E. Linden, *The education of Koko* (Holt, Rinehart, Winston, NY, 1981).
5. T. A. Sebeok and D. J. Umiker-Sebeok (Eds.), *Speaking of apes: A critical anthology of two-way communication with man* (Plenum Press, NY, 1980).
6. H. S. Terrace, Apes who "talk": Language or projection of language by their teachers?, in: *Language in primates: Perspectives and implications,* edited by J. de Luce and H. T. Wilder (Springer-Verlag, NY, 1983), pp. 19–42.
7. C. R. Thompson and R. M. Church, An explanation of the language of a chimpanzee. *Science*, **208**, 313–314 (1980).
8. J. L. Pate and D. M. Rumbaugh, The language-like behavior of Lana chimpanzee: Is it merely discrimination and paired-associate learning? *Animal Learning and Behavior*, **11**(1), 134–138 (1983).

CHAPTER SIXTEEN

Where do We Stand and Where Are We Going?

THE RESEARCH RESULTS

Most of the scientific work on animal language has been crammed into the last 35 years. Others will be able to make a better assessment of this work after another 35 years have passed. Nevertheless, we are confident about some conclusions.

First, the results obtained during the past 35 years have astonished many people, in some cases even those who were initially very skeptical about animals' ability to comprehend, let alone produce, rudimentary human language. Prior to 1966, there was no rigorous demonstration that animals could understand *novel* utterances either in a natural human language or an artificial language of gestures. Now there are several such demonstrations, with common chimpanzees, bonobos, gorillas, orangutans, dolphins, and even a parrot. With the exception of Alex, the parrot, production of language lags behind comprehension, as it does in the development of human children, especially hearing children. We do not know the size of Alex's comprehension vocabulary, although his productive vocabulary has been carefully assessed.

Again with the exception of Alex, learning to produce vocal language has been much less successful with animals than the use of signs or various language prostheses like lexigrams and plastic symbols. Apes have produced no more than five discriminable words vocally, but individual apes have used at least 170 signs very reliably, and the gorilla, Koko, has used as many as 1,000 appropriately at least once.

APPLICATIONS TO HUMAN LANGUAGE

The procedures that have proved most effective for teaching animals to comprehend human-designed languages are strikingly similar to the ways in which human children acquire language. For example, Kanzi, Panbanisha, and Panzee, to name a few cases among the apes, learned to use lexigrams primarily by observing their use by others rather than by reinforcing random or cued correct usage of the lexigrams. Using a single food reinforcer to teach animals to name objects was unsuccessful, but following up lexigram use with the named food or other object referenced was successful. More recently, the 10-month-old chimpanzee, Ayumu, correctly matched a lexigram with the brown square it represented after watching researchers train his mother, Ai, to read the lexigrams. *Discover* magazine[1] listed Ayumu's feat among the 100 top science stories of 2001; Kanzi, as we have seen, showed a very similar ability, beginning in 1982. But the main point here is that apes appear to learn best if they are treated like human children and reared in a language-rich environment. Thus humans serve well as a model for apes in this respect. With the benefit of hindsight, we can see that ape language researchers might profitably have paid more attention to how human children learn language and less to how operant psychologists teach animals other behaviors.

Some new techniques have been discovered for teaching humans with language difficulties since Cathy Hayes tried in vain to get ideas from speech therapists that would help her chimpanzee, Viki, to speak. One technique is called melodic intonation therapy. Aphasic patients can understand singing and can communicate through song, although they are unable to communicate through speech. In a related development, Paula Tallal found that language-handicapped children could understand slowed speech, but not speech delivered at a normal pace. These two techniques might be used in the future to help animals comprehend human speech, thus extending the use of the human model for teaching animals.

The modeling also goes in the other direction, from animals that do not ordinarily acquire human languages to humans who have unusual difficulty in doing so. The techniques used to teach animals are useful in teaching handicapped humans to communicate better, and indeed nearly every special training technique or device that has been used with apes has demonstrated effectiveness with human children—lexigrams, plastic symbols, and sign language have all proved useful. Sign language had, of course, long been used to allow deaf children to communicate, but its potential for helping language-handicapped hearing children had largely gone unrecognized.

The same can be said for the technique used with the parrot, Alex. Irene Pepperberg adopted and modified the procedure developed by Todt, and provided Alex with opportunities to observe verbal exchanges between a speaker and listener, and to enter the conversation when he wished to do so. Her modification was that she had the speaker and listener exchange roles, which demonstrated that communication could occur in both directions. Clearly one species can learn from another, although it is likely that a close bond between teachers and students facilitates learning. Probably parrots and apes are both good models for humans, so the modeling goes in both directions!

It is difficult to overstate the benefit that even very limited language offers to handicapped children and their parents. It is heartwarming to see the smiles on children's faces when they realize that they have attained some control, however limited, over their social environment, and those smiles are reflected by their parents. Future improvements in the methods of teaching animals, it now appears, will be almost certain to produce improvements in methods of teaching language to language-handicapped humans.

Most financial support for research is justified because of the above kind of actual or potential benefit to humans. Animal language research has been no exception to this rule, but has come to rely less on direct application to humans as people have become more concerned with the welfare of, and interest in, animals per se. In addition, we humans seem to be waking up to the fact that human welfare is intimately connected to the welfare of the whole biosphere, so the indirect benefits of learning more about all animals are becoming clearer to us.

THE IMPLICATIONS OF THE RESEARCH RESULTS

Animal Intelligence

Although the empirical results and benefits have been fascinating, the larger implications are even more fascinating. First among them is the demonstration that the language gap between humans and other animals is smaller than expected. Several species have demonstrated that they have the cognitive preparation that makes rudimentary language possible. It was not so much of a shock to find that chimpanzees, who are generally thought to be about 98.5 percent genetically identical to humans (that estimate has recently been questioned), have that cognitive preparation. Even the dolphins' abilities were not a huge surprise; dolphins have long been regarded as intelligent, large-brained animals, and are mammals. Thus we can with reasonable ease accept the possibility that evolution has prepared such animals to learn language. But how about the parrot, an animal that split off from the evolutionary tree that led to humans many, many millions of years ago? The parrot was regarded as an empty-headed imitator of sounds. Dr. Irene Pepperberg showed that, with proper and extremely persistent training, it could become much more. Thus the animal language research has boosted our estimation of animal intelligence; animals have learned things that, until the 1970s, were thought to be beyond their reach.

When Kanzi was asked to "make the doggie bite the snake," and did just that, when first hearing that sentence, the notion that animals could learn only specific stimulus-response connections died forever. That misguided notion was already moribund because of the evidence that dolphins were able to respond to new word orders on first hearing them, and that other apes seemed to be sensitive to word orders. This sensitivity to syntax, however rudimentary it may be, could not help but increase our estimates of animal intelligence.

Animal Consciousness

Animal accomplishments have not only confirmed the continuity between humans and other animals that evolutionary theory said should be present, but also confirmed the claims of biologist-philosophers like Griffin,[2] who in 1976 (or earlier) was questioning the doctrine that animals lacked conscious awareness. When Koko signs that she is "fine animal gorilla" and puts on her lipstick with the aid of a mirror, we are hard pressed to deny her the full panoply of human feelings. Every species of ape trained in a human language has shown strong evidence of self-awareness, either via the widely accepted mirror test or via informal observations—usually both! Dolphins have recently provided similar evidence, and even Alex, the parrot, has given weaker indications of self-awareness. Given self-consciousness, it is illogical to deny consciousness to animals. The reverse of course is not true; animals could be conscious in a weaker sense without being self-conscious. Thus the proof of self-consciousness in some animals makes us more likely to accept consciousness in a wider range of animals than before—certainly to mammals and likely to "lower" forms like birds or even reptiles.

Animal Rights

Roger Fouts[3] commented on the status of animals, after spending most of his adult lifetime in the study of chimpanzees:

> Even some scientists who are supposed to know about the plight of apes have argued against giving any legal rights with which they could be protected. Frans de Waal, an apologist for the use of chimpanzees in bio-medical research, warned in an Op-Ed piece that we should beware of the "slippery slope" if we granted apes the basic legal rights as proposed by the Great Ape Project. De Waal fears if we do this then before long we might be expected to grant rights to rats. What he and others like him seem to be ignorant of is that the slope he is so afraid of sliding down is one we have never tried: namely, the slope of compassion. The slope he advocates staying on is one we have never left: the slope of exploitation. ... Maybe if we showed compassion for a rat, we might even take the radical step of extending it to a forest, or the whole planet. (p. 5)

People may agree with either de Waal or with Roger Fouts, but we believe that understanding animal language capacities cannot help but move them toward a more empathetic position, and, as we said in the preface, the slippery slope of compassion should extend upward to humans as well as downward to lower animals. In this time of terrorism and wars, civil and uncivil, and of general unrest, such upward and downward compassion seem to be needed more than ever.

Certainly compassion is needed for all the great apes. Their numbers have decreased at an alarming rate; two primary reasons are loss of habitat and poaching. The bush meat trade is a crime approaching human cannibalism in the extremity of its immorality. Every animal language researcher recognizes the different personalities of his or her subjects. What a great word PERSONality is in this context! Communication with an animal encourages, or forces, observers to recognize the personhood of animal subjects. So eating that animal, or others of his or her species, becomes an unthinkable horror. Efforts to pass along an extreme aversion to the people who have free-living apes as neighbors are often successful. It is difficult to view a videotape of a bonobo (for example) conversing with a human and then go into the forest to shoot another bonobo to sell as bush meat.

Thus, we believe the greatest benefit of studying animal language, or seeing language-trained animals on television or videotape, or reading about them, is that it forces the observer to recognize animals personalities, their near-humanness. Alex can talk better than we can fly, so perhaps he deserves protection equal to ours!

THE FUTURE OF ANIMAL LANGUAGE RESEARCH

Predictions are difficult. It would be safest for us to take a page from Nostradamus and make statements like "Under the sun in the year of the great ape a brilliant researcher will arise in the west, and her feats will force envious others to bow down before her." But such predictions are not very helpful; the few general guesses that we will make will not be much better, based, as most will be, on research that is already under way.

Exploring Limits on Acquiring Syntax

One line of research with apes will investigate the limits of syntactic under-standing that can be reached in apes and dolphins given special training and testing. We mentioned above that Kako[4] questioned animals' ability to learn to use closed-class items and respond to syntactical structure. Closed-class items are words in a language that tend to remain stable, rather than admit-ting new words. By contrast, open class words easily admit new members, like jet plane, astronaut, fedex, and software. Closed-class categories of words include demonstratives, prepositions, conjunctions, and grammatical markers like the *s* that in English connotes plural and the *ed* that indicates that an action occurred in the past.

Parrots and other animals may also be trained and tested, and, given Alex's ability to distinguish between different types of questions, it may be too early to say that they will be less competent learners of syntax. Herman and Uyeyama[5] took up Kako's challenge immediately, and argued that dolphins had already demonstrated comprehension of both closed-class items and argu-ment structure. Neither Pepperberg[6] nor Savage-Rumbaugh, Shanker, and Taylor[7] showed much enthusiasm for answering Kako's questions, arguing that continuing to respond to challenges about animals' language abilities was not productive. However, enthusiastic or not, this line of research is certain to be pursued, and its results, whether positive or negative, will be fascinating and will help to solve the puzzles about the cognitive underpinnings of human language.

New Ways to Circumvent the Vocal Channel

Researchers will continue to search for more effective methods to circumvent the vocal channel or to make more effective use of it. So far, two techniques have been most promising: the use of sign language, and the use of lexigram boards. Each has advantages and disadvantages. The use of sign requires no special equipment, and is, therefore, the ultimate in portability and convenience. However, it is difficult to be certain immediately about the meaning of a sign, and it is very time-consuming to record and interpret signs via film or videotape. In addition, the anatomy of apes and, especially, dolphins, limits the signs that can be made. This limitation is certainly related to body structure, and may also involve limitations on the neural control of body movements. Past researchers using sign have accepted modifications of signs related to these limitations, and future researchers will probably design systematic sets of signs that intentionally circumvent these limitations.

Computerized lexigram boards make it possible to record responses immediately and objectively, so that there is no problem connected with identifying the response. Further, keypresses can produce sounds, so that words in any language (usually English in the past) can be produced by the keypress. However, sound production requires that a computer be activated by the keypress, so portable lexigram boards that are simple plasticized cardboard lack this advantage. Future research on lexigram boards may use more iconic, easily learned keys, now that it is no longer necessary to demonstrate that apes can use arbitrary symbols. Lexigram boards will also be further developed for underwater use by dolphins, and modified for greater ease of use by marine animals. It seems unlikely that the plastic symbols used by the Premacks will undergo future development because keyboards appear to incorporate their advantages without incorporating their disadvantages.

New Approaches to Animal Vocal Language

It is possible, although perhaps less likely, that two methods for using vocal production will be examined in more detail. One speculative method would require prostheses to modify the ape vocal tract to make it more human-like;

the rationale would be that the apes' limitation lies in the anatomy of the vocal tract, rather than in the neural control of the tract. We regard this approach as unlikely to succeed because the evidence to date indicates that the limitation is in neuromuscular control, not in the vocal tract.

The second method would be to use a set of phonemes that apes can already produce to construct a limited artificial language. At least one attempt of this kind was abandoned rather quickly because the experimenters could not understand the resulting words. Although humans could, with sufficient effort, learn a language comprised of such a limited sound set, apes might be incapable of producing the sounds despite the theoretical capability of their vocal tracts to produce the sounds. However, success in this endeavor would be marvelous. Kanzi tries to mimic the sounds of speech, and indicates his frustration by pointing to his mouth and throat after unsuccessful attempts to imitate words. It would be wonderful to free him, or other apes, from this limitation, as some language-handicapped children have been freed! Dr. Sue Savage-Rumbaugh believes that Kanzi and Panbanisha are developing a Creole of English and "Bonobo"; if so, this approach starts to look feasible.

We have already seen that Lieberman thinks the chimpanzee vocal tract (and no doubt the bonobo) could support a complex language. Lyn Miles[8] believes that apes could be taught a verbal language of select sounds, but adds " ... a series of popping and guttural sounds would appear even less humanlike and less appealing than present non-vocal methods, and interest in such methods will probably remain low" (p. 391). However, it is possible, at least in theory, to have a computer map the "popping and guttural sounds" into corresponding human phonemes and produce them through a speech synthesizer, so that a great ape speaking into a microphone would be heard by a human wearing headphones as making the sounds to which he or she was accustomed. That would avoid the "unhumanlike" problem, but would be of no help if the animal simply could not use the vocal system to communicate. This approach will almost certainly be tried in the future, according to the axiom "anything that can be done will be done."

Connecting Natural Animal Communication to Human-designed Language

Natural animal communication, in particular bird song, has been studied for a long time. The languages of elephants and dolphins have received a lot of attention, and the warning cries of vervet monkeys and the dances of bees have become classic cases of animal communication. However, the study of animals' natural communication systems has been for the most part separate from the study of animals' acquisition of a human-designed language. Some investigators have pointed out similarities between natural gestures and the signs of American Sign Language; "gimme," for example, is similar in both. It is a safe bet that future researchers will have studied their participants' natural gestures very carefully before embarking on a training program, especially if they use signs or vocalizations that resemble the animals' natural vocalizations. Further, they will be aware of the contexts in which the natural communications are used, and will use the information to design better teaching techniques for the animals.

Studying the Cognitive Components of Language in Multiple Species

Now that Alex has demonstrated what he can do with his small brain, it is bound to occur to future researchers that animals with brain sizes between the parrots' and the apes' may, with appropriate training, acquire some rudimentary language. The results of this research will contribute to our knowledge of the evolutionary pathway that leads from stimulus-response learning through the ability to recognize semantic relationships to the ability to acquire some rudiments of syntax and finally to human language. If we can unravel this evolutionary pathway, we may finally come to understand what the "language acquisition device" that Chomsky postulates accomplishes for us. That is, we might understand what cognitive processes are involved, once we unravel what evolutionary, cognitive, steps lead from less accomplished to more accomplished communicators.

THE MULTIMODAL APE

I think it might be possible to create what I will call a multimodal ape. She would be created by combining the existing techniques in a training regimen that would begin soon after birth. This ape would be reared in an environment combining intimate contact with other apes and with humans, much as Dr. Sue Savage-Rumbaugh does now. The ape would have contact with a computer that reinforced different vocalizations differently and allowed the infant ape student to control its environment via vocalizations. Training in gestures would begin very early, and would be based on all existing knowledge of the species' natural gestures, so that social exchanges between the ape student and researchers could use these gestures and build on them to create additional gestures. Meanwhile, lexigrams would be available to humans and apes, but the lexigrams would be as iconic as they could be made; some words cannot, of course, be made completely iconic, but words like above, beside, in, etc. can be visualized to some extent. Model-rival training would be used where appropriate. Vocal, sign, and lexigram communication all would be encouraged throughout; Dr. Savage-Rumbaugh already does many of these things, but the addition of iconicity, the model-rival approach, and training in syntax might be important additions to her methods. I would love to see a 10-year-old multimodal ape after this training regimen!

FUNDING THE MULTIMODAL APE PROJECT

If the past predicts the future, it may be very difficult to find money to finance a project that should last 10 years at a minimum. You have seen that Allen Gardner, Herbert Terrace, Lyn Miles, and Penny Patterson faced daunting problems in getting the support they needed, and Terrace dropped out of animal language research with Nim after only 3 years. Roger and Deborah Fouts almost literally had to go dumpster diving to feed their chimpanzees, as they went through Albertson's aging fruits and vegetables. Funding agencies of the United States Government have seldom been willing to commit themselves to support a line of research for 10 years, and animal language research has apparently not been a high priority.

Part of the reason for this is that animal language research has not been nearly as well known as it should have been. Many people have never heard of it, and are astonished to learn that apes have mastered a considerable number of signs. I will, however, hazard a prediction that the future will be rosier than the past has been for animal language researchers. People are increasingly concerned for the environment and our fellow travelers in it, and they are more interested in animals than ever before. As Tiny Tim said, "God bless us every one."

REFERENCES

1. C. Rist, A chimp off the old block. *Discover*, **23**(1), 69 (2002a).

2. D. R. Griffin, *The question of animal awareness* (The Rockefeller University Press, NY, 1981).

3. R. Fouts, The state of the planet of the apes. *Friends of Washoe*, **22**(1), 3–5 (2001).

4. E. Kako, Response to Pepperberg; Herman and Uyeyama; and Shanker, Savage-Rumbaugh, and Taylor. *Animal Learning & Behavior*, **27**(1), 26–27 (1999).

5. L. M. Herman and R. K. Uyeyama, The dolphin's grammatical competency: Comments on Kako. *Animal Learning & Behavior*, **27**(1), 18–23 (1999).

6. I. Pepperberg, Rethinking syntax: A commentary on E. Kako's "Elements of syntax in the systems of three language-trained animals." *Animal learning and behavior*, **27**(1), 15–17 (1999).

7. S. Savage-Rumbaugh, S. G. Shanker, and T. Taylor, *Apes, language, and the human mind* (Oxford University Press, NY, 1998).

8. H. Lyn Miles, Anthropomorphism, apes, and language, in: *Anthropomorphism, anecdotes, and animals*, edited by R. W. Mitchell, N. S. Thompson, and H. L. Miles (State University of New York Press, Albany, 1997), pp. 383–404.

CAST OF CHARACTERS

HUMAN ADULTS

Madame Abreu. Resident of Cuba who maintained a colony of chimpanzees; Gua, the Kellogg's chimp, was born in her colony on November 15, 1930. She was later transferred to the Anthropoid Experiment Station of Yale University at Orange Park, Florida. Robert Yerkes allowed the Kelloggs to take Gua on June 26, 1931.

Elizabeth Bates. Linguist specializing in children's language acquisition; she is sympathetic and helpful to researchers in animal language.

Sarah Boysen (1949–). An early associate of Duane Rumbaugh and Sue Savage-Rumbaugh, she now directs a large chimpanzee project (11 animals) at Ohio State University, and has obtained fascinating data on chimpanzees' ability to count. They can choose the smaller of two Arabic numerals in order to get a larger portion, but when the concrete food objects are presented, the animals cannot resist choosing the larger portion, and hence receive the smaller.

Roger Brown (1925–1997). Student of child language; emphasized differences between child and chimpanzee acquisition of language.

Janis Carter (19??–). A student at the University of Oklahoma, she became caretaker of the Temerlins' chimpanzee Lucy, and spent harrowing years trying to reintroduce Lucy to the wild in The Gambia.

Noam Chomsky (1928–). Famous linguist, proponent of linguistic theory involving innate language acquisition devices, and of a transformational grammar that expresses the "deep structure" of thought in the "surface structure" of language, and vice versa; he minimizes both ape language ability and the significance of language research with apes.

Ronald Cohn (1943–). An electron microscopist and recombinant DNA specialist, he is a long-time companion of Penny Patterson and the gorillas Koko, Michael, and Ndume.

Michael Crichton. Author of *Congo*, a novel in which a gorilla trained in sign language serves as an interpreter of the language of a band of renegade gorillas to an expedition of humans. The possibility of such interpretation had been suggested semi-seriously by Duane Rumbaugh many years ago.

Deborah Fouts (1943–). Currently at Central Washington University, she is a language researcher and co-keeper (with husband, Roger Fouts) of four chimpanzees: Washoe, her adopted son, Loulis, Tatu, and Dar.

Roger Fouts (1943–). Also at Central Washington University, he is a language researcher and co-keeper (with Deborah Fouts) of the four chimpanzees. He was a student of the Gardners at Reno, and was later at Oklahoma University and the Institute for Primate Studies directed by William Lemmon before moving to Washington. According to Eugene Linden, Fouts left literally in the dead of night, perhaps because of a controversy about whether Washoe belonged to Fouts or to Lemmon.

William Henry Furness 3rd (1866–1920). Furness was a physician/ surgeon who tried, beginning in 1909, to train orangutans to speak; one learned to say "Papa" and "cup," and apparently understood that the former was to be used as Furness's name. The female orang also learned to say "th"; Furness intended to use this sound as the basis for "the," "this," and "that," but his student died before his plan could be completed. Another orangutan "refused to be educated."

Allen Gardner (1930–). Psychologist, Ph.D. from Northwestern University. With his wife Beatrix, Dr. Gardner made the first known systematic effort to teach sign language to a chimpanzee. Their first student was Washoe, who mastered 132 signs to the conservative criterion used by the Gardners. Other chimps they studied later (Moja, Tatu, and Dar) showed similar competence.

Beatrix Gardner (1933–1995). Beatrix trained as an ethologist at Oxford under Niko Tinbergen. With her husband Allen, she made the first known systematic effort to teach sign language to a chimpanzee. She died suddenly on June 5, 1995 during a triumphal tour of Italy with Allen. Robert Solso in her obituary reports an unbelievably poignant irony; near the end, her jaws were locked by the septicemic organism that killed her, and she conveyed her last thoughts to Allen in American Sign Language, which she likely would not have known except for the Washoe work.

Richard Lynch Garner (1848–1920). In the summer of 1893, Garner spent 112 days in a cage in Gabon so that he would be protected while studying chimpanzees and gorillas in their native habitat. Garner may have been the first person to teach a chimpanzee to vocalize a word; he claimed that his chimpanzee, Moses, spoke a good rendition of the French word, feu, which is translated in English as "fire," and inferior versions of the German "wie" (how) and the nearly universal "mama," for which he made the lip movements but not the sound.

Jane Goodall (1934–). Jane Goodall is the best-known observer of chimpanzees in the wild, and is the author of many books and articles. She has followed the chimps in Gombe Stream Reserve for about 30 years, and is a committed and highly effective protector of primates.

Catherine Hayes. Viki's "mama" is the author of *The Ape in our House*, which describes adventures with Viki in an unpretentious and popular style. She is chided by Candland (1993) for losing scientific perspective, but the present authors cannot find fault with anyone who suffered through the life and death of a young chimpanzee.

Keith Hayes (1921–). Viki's "papa," who with Cathy Hayes raised Viki in an environment as nearly like that of a human child as possible, from near birth until Viki died of encephalitis at the age of seven.

Louis Herman (1930–). Herman is a marine mammal expert whose Ph.D. is from Penn State, now at the University of Hawaii. He has taught dolphins to respond to visual or vocal signs, and believes that they are responding to syntactical relationships.

Maria Hoyt. Mrs. Hoyt was obviously an upper crust woman of the depression whose husband retired upon their marriage, and who could afford to buy and maintain a gorilla and adapt her life style to that of the gorilla (named Toto). Although Toto was not given special training, Ms. Hoyt claimed that Toto understood as much Spanish as did a child of the same age. Toto, however, was not tested systematically.

Louise Kellogg. She was the wife and colleague of Winthrop Kellogg, mother of their child Donald, and adoptive mother for 9 months of the chimpanzee Gua.

Winthrop Kellogg (1898–1972).   Kellogg was a psychologist from Indiana University who observed his child Donald being reared for 9 months with the chimpanzee Gua, and reported (with his wife, Louise) the results in a book and other publications.

Shigeru Kiritani (19??–19??).   Kiritani is a contemporary Japanese scientist who publishes on the hearing and vocalizations of chimpanzees.

Nadesha Kohts (189?. –19??).   Kohts was a Russian woman who compared the development of a chimpanzee, Joni, during the years 1913–1916 with that of her son, Roody, during the years 1925–1929.

Shozo Kojima (19??–19??).   Kojima is a contemporary Japanese scientist who publishes on the hearing and vocalizations of chimpanzees.

Karl Krall (1861–19??).   Krall was a horse trainer who took over ownership of Clever Hans II from Herr von Osten, and trained Muhamed and Zarif in similar performances.

William Lemmon (1916–1986).   Lemmon's Ph.D. in clinical psychology was from Ohio State University. He headed the Institute for Primate Research at the University of Oklahoma. He was a large, bald man with an unusual ability to dominate chimpanzees, and, if Linden, Temerlin, and Roger Fouts can be believed, even humans. He feuded with Roger Fouts over the treatment of chimpanzees.

Philip Lieberman (1934–).   Leiberman is a, perhaps the, leading expert on vocal tract analysis. He has published extensively on the comparative analysis of the vocal tracts of chimpanzees, Neanderthals, and modern humans. His computer analyses indicate that chimpanzee vocal tracts could produce many, but not all, of the sounds of human language, enough, at any event, to support a complex language. His analyses indicate that much of the difficulty with teaching chimpanzees vocal language lies in neural and/or muscular factors rather than solely in the anatomy of the vocal tract.

John Lilly (1915–2001).   In the 1960s Lilly made extravagant claims about dolphins' ability to understand and produce human language. His claims are generally viewed with skepticism, and he acknowledged that he was unable to communicate directly with the dolphins. Schusterman prefers to explain the dolphins' and seals' performances in terms of conditional discriminations and/or simple rule-following rather than in terms of linguistic ability.

Eugene Linden.   Linden is the author of several books on ape language. His books are based on personal contact with several of the researchers.

Hugh Lofting (1886–1948).   Lofting was an English author and illustrator; he wrote the story of Dr. Dolittle, the fictional character who could talk to animals.

James Mahoney (19??–).   Mahoney is a veterinarian and was Associate Director of the Laboratory for Experimental Medicine and Surgery in Primates (LEMSIP). He is respected by those on both sides of the animal rights controversy, and is sympathetic to the plight of the chimpanzees in his care, some of whom were trained in sign language.

Julien Offray de la Mettrie (1709–1751).   La Mettrie was a French philosopher who suggested that apes should be able to learn a vocal or sign language.

Lyn Miles (1944–).   Dr. Miles taught Chantek, an orangutan, "pidgin sign," beginning in 1978.

Jan Moor-Jankowski.   Dr. Moor-Jankowski was Director of the Laboratory for Experimental Medicine and Surgery (LEMSIP) from 1965–1995. He is highly principled and sees both sides of animal rights issues. He was recently discharged as director after blowing the whistle on a researcher who was conducting drug addiction studies in rhesus macaques by forcing them to inhale a cocaine mist.

C. (Conwy) Lloyd Morgan (1852–1936).   Lloyd Morgan was an English comparative biologist whose canon stipulates that a behavior should not be explained by a psychic function higher in the psychological scale if a function lower on the scale is sufficient; this principle is usually, though perhaps mistakenly, regarded as an adaptation of Occam's razor.

Horapollo Nilous (ca. 500 A.D.).   Nilous was an Egyptian scribe who reported that priests from ancient times tested baboons newly arrived at the temple for their ability to read and write. Their test appears to be the first formal evaluation of the ability of animals to understand human language.

William of Occam (also spelled Ockham) (c. 1280–1349).   Occam stated a principle called Occam's razor, also called the principle of parsimony, which forbids the multiplication of unnecessary entities in explanations. The simplest adequate explanation of any observation is to be preferred.

Penny Patterson (1947–).   Dr. Patterson received her Ph.D. in psychology from Stanford University based on her work with Koko the gorilla. She has been a devoted trainer of Koko and of Michael until he died. With Ron Cohn, she now cares for Koko and Ndume in Woodside, California, pending a move to better quarters on Maui.

Irene Pepperberg (1949–).   Dr. Pepperberg earned a Ph.D. in theoretical chemistry from the Massachusetts Institute of Technology, but she soon turned to teaching Alex, an African Gray parrot, to talk. She was for a time at Northwestern University, then at the University of Arizona, and now back at MIT. Alex can perform a number of cognitive tasks, including naming several shapes and colors, and respond in English. Alex's vocal productive vocabulary probably exceeds that of any other non-human animal.

Samuel Pepys (1633–1703).   Pepys was an English author whose diary is extremely well known. He saw a "great baboone" on the docks in England on August 21, 1661, and suggested in his diary of that day that it might be taught to speak or make signs.

Oskar Pfungst (1874–1932).   Pfungst was the careful student of Carl Stumpf who was assigned to Clever Hans and who demonstrated that Hans could not give correct answers unless a human was present who knew the correct answers. Hans' apparent ability to take advantage of cues given by human observers led to calling this use of cues the "Clever Hans effect" and to criticism of any performances that might be based on such effects.

Steven Pinker (1954–)   Pinker is a "Chomskyian" linguist who denigrates the study of animal language as a highly flawed enterprise. He is author of *The Language Instinct*, an excellent book that, as its title suggests, espouses a nativist view of human language.

Ann Premack (1929–).   She is half of the Premack team that has been studying chimpanzees' use of plastic symbols for words for about 40 years. She is the writer of *Why Chimpanzees Can Write*, and coauthor of several other publications. She, like her husband David, has star quality, but she underplays it.

David Premack (1925–).   David Premack's Ph.D. is from the University of Minnesota. He is often regarded as brilliant, charismatic, and a little eccentric. He proposed the principle of reinforcement that became known as the Premack principle, in addition to working as a pioneer in the study of chimpanzee language with Sarah as his star pupil and Ann Premack as his primary collaborator.

Duane Rumbaugh (1929–). Rumbaugh started research with animals at the San Diego Zoo soon after earning his Ph.D. from the University of Colorado. He was one of the very early animal language researchers, and became the Director of the Language Research Center in Decatur, GA, and of the LANA project with the chimpanzee Lana. He and von Glasersfeld developed the Yerkish language, and he was the first researcher to use a computer for teaching language to chimpanzees. He has since collaborated with Sue Savage-Rumbaugh on later work with Austin, Sherman, Kanzi, Panzee, Panbanisha, and other apes.

Sue Savage-Rumbaugh (1946–). Dr. Savage-Rumbaugh is an innovative language researcher who first worked with chimpanzees at the University of Oklahoma while earning her Ph.D. She is the trainer and companion of Kanzi and other bonobos who are probably the most language-sophisticated of all animals. She is a careful, solid, and prolific researcher and publisher, devoted both to science and to her bonobos. She has produced the most careful analysis and empirical investigation of semanticity in animal language. She is also deeply engaged in the study and preservation of bonobos in captivity and in their native habitats.

Ronald Schusterman (1932–). Dr. Schusterman is a biological psychologist whose Ph.D. is from Florida State University; he is now associated with California State University, Hayward. He trained the sea lion Rocky to respond to arbitrary signs, and claims that Rocky's linguistic competence is comparable to that of dolphins or chimpanzees. He prefers to explain the animals' performances as conditional sequential discriminations and/or as simple rule following rather than by using grammatical terms.

Thomas Sebeok (1920–2001). Sebeok was a linguist at Indiana University; with Jean Umiker-Sebeok, he edited a book for which he wrote an introduction highly critical of, if not derisive about, the reports of animal language competence. He was at least as guilty of unsupported criticism as his victims were of unsupported claims.

Burrhus Fred Skinner (1904–1990). Perhaps the most famous psychologist of all time, Skinner was a persistent student of operant behavior, and generalized operant explanations to all organismic behavior. His attempt to apply an operant analysis to language in his book of 1956, *Verbal Behavior*, elicited an extremely hostile criticism from Noam Chomsky. However, Skinner's extremely broad definition of verbal behavior as "any behavior which requires the intervention of another organism for its reinforcement" may have made it easier for investigators like the Gardners and the Premacks to consider bypassing the vocal channel in their animal language research.

Maurice Temerlin (19??–).   Dr. Temerlin was a patient and colleague of William Lemmon, "stepfather" of chimpanzee Lucy, and the author of *Growing Up Human*. Lucy was later returned to the wild through the incredible dedication of Janis Carter.

Jane Temerlin (19??–).   Jane Temerlin was Lucy's "stepmother" and Maurice Temerlin's wife.

Herbert Terrace (1936–).   Terrace is a Skinnerian psychologist who earned his Ph.D. at Harvard University. He was responsible for the chimpanzee Nim's training in sign language. Although he was initially very positive about teaching Nim through operant procedures, he concluded that chimpanzees could not construct sentences. His criticism was probably overstated and then greatly overinterpreted, and probably contributed to a downturn in interest in research in animal language. He implied at times that chimpanzees could only imitate, but his written criticisms were more limited and quantitative, and in the book, *Nim*, he strongly encouraged further research and recognized the limitations of his own study with Nim.

Jean Umiker-Sebeok (19 –).   Dr. Umiker-Sebeok was the co-editor (with Thomas Sebeok) of the highly critical book on animal language research.

Herr von Osten (1838–1909).   Von Osten was the owner/trainer of Clever Hans, and the unwitting creator of the "Clever Hans effect." He was most likely an honest man who was, as he claimed, deceived by Hans.

David Washburn (1961–).   Dr. Washburn is a psychologist whose career has centered on Georgia State University; he is a colleague of Duane Rumbaugh, Sue Savage-Rumbaugh, Sherman, and Austin at the Language Research Center, where he is now director. He is the author or co-author of several papers on animal language, counting, and perceptual-motor performance.

Robert Yerkes (1876–1956).   Yerkes was both a researcher in primate behavior and an administrator after whom several laboratories of primate biology are named. He was one of several pioneers who suggested that vocal language might be impossible for chimpanzees, while sign language remained a possibility.

CHILDREN

Alia (July 26, 1987–).   Alia is the daughter of Jeannine Murphy, who was a colleague of Sue Savage-Rumbaugh. Alia's comprehension of English sentences was compared to Kanzi's, beginning when she was 2 years old and Kanzi was a little over 7 1/2. They were almost equally competent.

Donald (1930–).   Donald was (is?) the son of Winthrop and Louise Kellogg, raised for 9 months with the female chimpanzee, Gua, who was 2 1/2 months younger than Donald.

Roody (1925–).   Roody was Nadesha Kohts' son; she compared his behavior of 1925–1929 to that of her chimpanzee Joni from 1913–1916.

CHIMPANZEES

Aaron (1892–189?).   Aaron, a kulu-kamba, was one of Richard Lynch Garner's favorite acquisitions. Garner contrasted Aaron's wonderful disposition with that of the female Elisheba, another kulu-kamba who enjoyed nothing so much as irritating Aaron, who in turn treated her in a most caring manner.

Ally (1969–).   Ally is or was a language-trained chimpanzee raised for 4 years in a home, then returned to the Institute for Primate Studies in Oklahoma where he was born, thence briefly to LEMSIP, where he became a *cause celebre*. Ally was then returned to Oklahoma, largely forgotten, and probably then shipped as a breeder to White Sands, New Mexico. According to Linden and to Fouts, he has probably been renamed (or un-named) and left with his signing ability unknown and unused.

Austin (1974–1996).   Austin became Sherman's companion at the Language Research Center in Decatur, Georgia, and participated with him in the groundbreaking interchimp communication experiments. He remained there until his death in 1996 of causes that could not be ascertained at autopsy.

Bobby (1986–).   A participant in Sarah Boysen's counting experiments at Ohio State University.

Booee (1967–).   Booee was born at a National Institutes of Health laboratory near Washington, D.C., and later transferred to Oklahoma, where he trained for a time on sign language with Roger Fouts. He resided at LEMSIP at last report. Booee caught Jim Mahoney in a lie when the latter denied having candy (because he didn't have enough for all chimpanzees present). Booee signed "Booee see sweet in pocket."

Bruno (1968–).   Bruno was Booee's companion at Oklahoma, where Bruno was born, and part of the Fouts troupe learning sign language at Lemmons' Institute. He was later sent to LEMSIP.

Dar (1976–).   Dar's full name is Dar es Salaam , for the capital of Tanzania. He was born on August 2 at Holloman Air Force base in New Mexico, and quickly adopted by the Gardners as one of their language students. Later the Foutses took over his care, and he is still with them in Ellensburg, Washington.

Erika (ca. 1972–).   Erika was a wild-born chimpanzee who, at the age of about 2 1/2, was assigned to Sue Savage-Rumbaugh's Animal Model Project, with Sherman, Austin, and Kenton. After 18 months the demands of working with four animals proved too much for the staff, and only Sherman and Austin continued in training.

Gua (1930–????).   Gua was a female chimpanzee born in Madame Abreu's colony on November 15, 1930, and raised for 9 months with their child, Donald, by Winthrop and Louise Kellogg, beginning on June 26, 1931. Gua was 2 1/2 months younger than Donald. At the end of the experiment, Gua was, according to the Kelloggs "returned by a gradual habituating process to the more restricted life of the Experiment Station".

Joni (1912–1917).   Male chimpanzee kept by Nadesha Kohts from 1913 to 1916, when Joni was 1 1/2 to 4 1/2 years old.

Lana (1970–).   Lana was named by the then director of the Yerkes Primate Center, Geoffrey H. Bourne. She became adept at using strings of key presses on a computer keyboard with lexigrams on the keys; the computer analyzed the strings for grammaticity, and delivered various rewards for correct strings. As might be expected on such a regime, Lana was more adept as a producer of lexigram strings than as a comprehender of the strings. She remains at the Language Research Center.

Loulis (1978–). His full name is Loulis Yerkes, after his place of birth on May 10. He was adopted by Washoe when he was 10 months old after Washoe's infant, Sequoyah, died at the age of about 2 months. Loulis learned 55 signs from other chimpanzees, and thus first demonstrated that chimpanzees could pass along a culture of sign language. He is still with Washoe, the other chimpanzees, and the Foutses at Central Washington University.

Lucy (1964–1988). Lucy was a chimpanzee raised as a human and trained in sign language by the Temerlins and researchers at the University of Oklahoma. She was returned to the wilds in The Gambia at the age of 13 by Janis Carter. She adapted after several years, but was killed by poachers.

Moja (1972–). Moja's full name is Moja LEMSIP, after her place of birth at the LEMSIP (Laboratory for Experimental Medicine and Surgery in Primates) lab. One of the chimpanzees with whom the Gardners worked, beginning when the chimpanzees were very young infants, she was later placed in the care of the Foutses, where she remained until her sudden death in 2002.

Moses (1892–????). Moses was Richard Lynch Garner's favorite common chimpanzee, who, Garner claimed, could say "feu" and make several interpretable chimpanzee vocalizations.

Nim (1973–2000). Nim was the famous chimpanzee jokingly named "Nim Chimsky" by Herbert Terrace as a takeoff on the name of linguist Noam Chomsky. Nim was returned to the Institute of Primate Studies in Oklahoma and later, following considerable controversy, placed on a ranch in Texas.

Panzee (1985–). Panzee is a chimpanzee who was reared and compared with Panbanisha at the Language Research Center. Panzee did not begin using lexigrams as soon as Panbanisha, but Panzee did start to use them by the time she was 18 months old. She has showed a remarkable ability to direct blind observers to food hidden several hours or even days earlier, via the use of lexigrams and pointing.

Pili (1973–1975). One of the Gardner's chimps; Pili died very young of leukemia.

Sarah (1957–).   Sarah was the Premacks' star pupil; she constructed vertically arranged columns of plastic symbols that were regarded as sentences. She was with Sarah Boysen at Ohio State University as of July, 2003.

Sheba (1981–).   Sheba is probably Sarah Boysen's star pupil in arithmetic at Ohio State University. She recognizes several Arabic numerals and can match them to displayed numbers of objects. She can also choose the correct numeral to represent the sum of small numbers of items placed in different locations.

Sherman (1973–).   Sherman was born in and remains in Georgia at the Language Research Center. Sherman and Austin were the first to demonstrate that chimpanzees could communicate with each other to solve problems that could not be solved without the use of symbols.

Tatu (1975–).   Her full name is Oklahoma Tatu, after her place of birth on December 30. She became one of the several sign language students with the Gardners who are now in the care of the Foutses. Tatu loves black, enjoys pointing out and labeling black objects and photographs, and is likely to label anything good "black."

Viki (1947–1954).   Viki was the most famous of the home-reared chimpanzees, both because, as Winthrop Kellogg said, she represented "the acme of chimpanzee performance" with her four words of spoken English, and because Catherine Hayes' book, *The Ape in Our House*, enjoyed such popularity.

Washoe (1966–).   Washoe was the first chimpanzee to be trained in the use of sign language, achieving a vocabulary conservatively estimated to be 132 or 133 signs while in Reno, and later at 170 signs while in the care of the Foutses. She probably understands many more signs. Washoe was wild-caught, moved to Reno, Nevada (which is in Washoe County) to be with the Gardners, and later accompanied Roger Fouts to Oklahoma and to Ellensburg, Washington, where she now resides. Washoe meant "people" to the Indian inhabitants of Washoe County, Nevada.

BONOBOS

Kanzi (1980–).   Kanzi has demonstrated more understanding of spoken English under controlled conditions that any other non-human animal in history. He comprehended almost exactly 3/4 of over 600 sentences that he had never heard before, a percentage slightly higher than that achieved by the 2 1/2-year-old child, Alia, with whom he was compared. Kanzi was born at the Yerkes laboratory in Atlanta, but has spent most of his life with Sue Savage-Rumbaugh at the Language Research Laboratory in the Atlanta suburb of Decatur.

Matata (1970–).   Matata is Kanzi's stepmother, who stole Kanzi from his real mother, Lorel, soon after Kanzi was born. Matata resides at the Language Research Center in Decatur, GA. Because she was caught in the wild, her exact date of birth is unknown.

Mulika (1983–).   Mulika, like Panbanisha, a half sister of Kanzi's, was given early exposure to the use of lexigrams and began using them by the age of 1 year.

Nyota (1998–).   Nyota is the son of P-Suke and Panbanisha; he has been reared largely by humans from birth, and so far seems ahead of Kanzi in pre-language development.

Panbanisha (1985–).   Panbanisha is Kanzi's half sister and is, like Kanzi, a resident at the Language Research Center, where she is trained in the use of lexigrams. Her ability is comparable to Kanzi's.

P-Suke (1975?–).   P-Suke is a bonobo given by the Japanese to the Language Research Center in 1996 for breeding purposes. Although he was reared with no bonobo companions, he has succeeded in introducing his new genes into the bonobo population quite well, having already sired infants with Matata and Panbanisha. His name is an attempt to mimic a peculiar cry that he habitually made.

Tamuli (1987–2000).   Tamuli is another daughter of Matata. Tamuli had little exposure to human language, and accordingly knew none at the time of her death of heart failure in 2000.

GORILLAS

Koko (1971–).   Koko (full name Hanabi Ko, for "fireworks child"), Francine (Penny) Patterson's famous gorilla, was born on July 4, hence her full name. She started her sign language lessons when she was about a year old. Penny, unlike the Gardners, did not forbid the use of vocal English in Koko's presence, and Koko has considerable English comprehension, as well as the ability to produce and understand signs.

Michael (1973–2000). Michael was a male gorilla with whom it was originally hoped that Koko would mate. That did not work; however, Michael was trained in sign along with Koko, and gained considerable fame for his artistic work with acrylics.

Othello (1891–1892). Othello was a gorilla studied briefly by Richard Lynch Garner. Othello was about 1 year old when Garner obtained him. He died a short time after; Garner believed that his native boy had poisoned Othello, who immediately before his death had seemed quite healthy.

Toto (193?–????). Toto was Mrs. Maria Hoyt's gorilla, who, like Koko, had a kitten and understood, according to her mistress's report, considerable spoken language. When she became too strong to keep around the house (she once broke Mrs. Hoyt's wrists in a fit of enthusiasm, and upon another occasion smashed the family station wagon into the garage wall, although the brakes were set), she was sent to become the mate for the famous Gargantua.

ORANGUTANS

Chantek (1977–). Chantek is the male orangutan engaged in sign language training by Dr. H. Lyn White Miles. Chantek was born at the Yerkes Regional Primate Research Center in Atlanta and trained for the most part at the University of Tennessee at Chatanooga, beginning when he was 9 months old. He was returned to Yerkes in 1986, but later moved to his present location at the Atlanta Zoo. He learned 140 signs and invented others of his own.

Unknown Soldier (1908–1912). This unnamed orangutan was kept by Furness for 4 years and 8 months. He was the first orangutan, so far as we know, to utter a word of human language.

DOLPHINS

Akeakamai (1977–). "Ake" was trained in a visual language in which words were represented by large-scale gestures of the hands and arms. The principles were similar to those used for sign languages used with apes and deaf humans.

Keakiko (197?–1977?). With Puka, "Kea" was one of the first dolphins entered in Lou Herman's language training course; she learned to respond to the nine possible sentences made up of three nouns and three verbs before being stolen and returned to the sea, where she probably died.

Puka (197?–1977?). Puka was the second of Herman's original dolphins; like Kea, she was stolen and returned to the ocean, which was probably a death sentence for both. Puka was not as advanced in her language training as Kea.

Phoenix (1977–). Phoenix learned to respond to an underwater, computer-generated set of acoustic signs. Her name symbolized the rebirth of Herman's laboratory after the theft of his original dolphins.

HORSES

Clever Hans I (189?–190?). This Hans was the first horse of Herr von Osten; little is known about Hans I.

Clever Hans II (190?–191?). This was Herr von Osten's second horse, the one who deceived him and the one who put Oskar Pfungst on the trail of the Clever Hans effect. Hans was evidently responding to cues unconsciously presented by humans in Hans' view who knew the correct answers to the questions given to Hans.

Muhamed (190?–19??). Muhamed was a horse owned by Karl Krall, who took over the care of Clever Hans II. Muhamed's feats paralleled those of Clever Hans.

Lady Wonder (1924–1953). Lady Wonder could allegedly spell words and even exercise telepathy, for example in finding two missing children in 1953.

Zarif (190?–19??). Zarif was another talented horse owned by Karl Krall. Both Muhamed and Zarif were said to be able to spell out words on a letter board constructed by Krall.

DOGS

Don (19??–19??). Don was owned by Fraulein Ebers early in the 20th century. He was said to be able to speak six words that were "intelligible to nonfamily members." The words were Don (which sounded like "ong"), haben, Kuchen, Hunger, Ruhe, ja, and nein. Don could not speak on rainy days. Whether Fraulein Ebers was a ventroloquist, or whether Don understood whereof he spoke, we do not know.

Fellow (1924?–19??). Fellow responded to over 200 English vocal commands. He was tested in 1928 by the psychologists Warden and Warner; the tests indicated that some of Fellow's performances were based on the Clever Hans type of cue (the giver of commands had to be present), but in about 100 other cases Fellow obeyed the vocal command correctly with the owner behind a screen. Warden and Warren thought it unlikely that Fellow's performance indicated true understanding of words, but thought him remarkable at discriminating different sounds.

PARROTS

Alex (1976–). Alex is an African Gray parrot (*Psittacus erithacus*) who has a remarkable ability to use English words meaningfully (referentially). Alex was trained, beginning at the age of 13 months, by Irene Pepperberg, and later by her students as well.

Griffin (198?–). Another of Pepperberg's gray parrots, Griffin has been used primarily as a subject in experiments designed to isolate the critical components of language training.

Kyaro (19??–). Kyaro is a third Pepperberg parrot, used in a way similar to Griffin.

SEA LIONS

Gertie (19??–). Gertie is a female sea lion, trained to carry out actions specified by the same kinds of visual signs used by Herman with dolphins.

Rocky (19??–). Rocky is the most accomplished of Schusterman's sea lions. Rocky was trained to respond to large-scale signs similar to those used by Herman with his dolphins.

INDEX

Aaron (kulu-kamba), 1, 22, 48, 49–50, 173
Abreu Colony, Cuba, 57
Acoustic information processing, in dolphins, 121, 216
Acoustic language, of dolphins, 221–224
Aesop, fables of, 1, 45–46
Age Language: From Conditioned Response to Symbol (Savage-Rumbaugh), 32
Aggression, in chimpanzees, 63
Ai (chimpanzee), 22, 162, 201–202, 270
  Arabic numeral use by, 205–206
  lexigram use by, 205
Ai Project, 201–211
  historical background and rationale of, 202–205
Akeakamai (dolphin), 214, 219, 220, 221–222, 223–224, 229, 230, 232
Akili (bonobo), 6, 174
Akira (chimpanzee), 202
Alarm calls. *See* Warning calls
Alex (parrot), 11, 193, 237–253
  birth, 5
  brain size, 241
  category identification ability, 242–245, 251
  courtship behavior, 238–239, 241
  English-language comprehension ability, 28
  intelligence, 238, 241, 251–252
  self-awareness in, 272
  syntax use by, 252, 274
  vocabulary of, 28, 245, 252, 269
  vocalizations, 239–240, 249
  vocal training of, 246
    comparison with training of chimpanzees, 247
    modeling in, 270–271
    model/rival approach in, 247–250
Alex Studies (Pepperberg), 241
Ally (chimpanzee), 66, 89, 189, 196
Altruistic behavior, in bees, 26
American Sign Language. *See* Sign language
Anatomy, dolphins' understanding of, 224
Animal activists, 218–219. *See also* Animal rights

Animal language
  early reports about, 45–67
    Clever Hans, 1, 2, 50–53
    in mythology, 1, 45–46
    Richard Lynch Garner, 1, 16, 22, 29, 47–50, 173
    William Henry Furness 3rd, 53–55
  interpretation of, 32–41, 148–149
  meaning in, 32–35
  overview of, 9–24
Animal language research. *See also* Gestures, communicative; Lexigrams; Plastic symbol language; Sign language
  with apes
    "Clever Hans" effect in, 255, 256
    early investigations, 47–50, 53–57
    evaluations of, 255–267
  application to human language, 267–271
  choice of species for, 27–28
  chronology of events in, 1–7
  difficulties in, 30–32
  future of, 274–277
  implications of, 271–274
  rationale for, 21–23
  response probabilities in, 226–228
Animal rights, 95–96, 194, 273–274
Animals
  ability to learn human language, 28–29
  human's changing perspectives on, 42–43, 110
Ape Language: From Conditioned Response to Symbol (Savage-Rumbaugh), 166
Apes. *See also* Bonobos; Chimpanzees; Gorillas; Kulu-kambas; Orangutans
  as food source for humans, 42, 273
  intelligence, 199
  lack of human language capacity, 136–137
  language research with
    "Clever Hans" effect in, 255, 256
    evaluations of, 255–267
    multimodal, 277–278
    verbal language ability, 10, 11, 269
Aphasia patients, song-based communication by, 270
Apple computers, 99–100

Tool production/use *contd.*
  intergenerational transmission of, 74, 208
  by orangutans, 192
Toto (gorilla), 31–32, 42, 55–57, 100
Towne, Robert, 89
True Friends, 95
Tursiops truncatus, encephalization quotient of,
  215

Umiker-Sebeok, Jean, 6, 52, 261, 266
United States Department of Agriculture, 95, 96
University of Arizona, 237
University of California, San Diego, 202
University of California, Santa Barbara, 114
University of Connecticut, 189
University of Georgia, 128, 129–130
University of Hawaii, 213
University of Missouri, 114
University of Nevada, Reno, 62, 87
University of Oklahoma, 154–155
  affiliation with Institute of Primate Studies,
    31, 50, 63, 64, 87–89, 189
University of Pennsylvania Medical School, 53
University of Tennessee, 189

Verbal language ability
  of apes, 10, 11, 269, 276
  of autistic children, 14–15
  of bonobos, 14, 152, 164, 165, 178–179
  of chimpanzees, 11–15, 16, 17, 48, 59–60,
    80, 152, 161
  of gorillas, 48
  of kulu-kambas, 49
  new approaches to, 275–276
  of nonhuman primates, 15–16
    neurological factors in, 12–13
  of orangutans, 54
  of parrots, 10, 15, 269
Videotaping, of chimpanzees' sign language use,
  80, 258
Viki (chimpanzee), 4, 11, 14, 30, 48, 58–59,
  59–61, 70, 73–74, 80, 161, 270
Vision, in dolphins, 215–216
Vocabulary
  of bonobos, 10
  of chimpanzees, 11, 48, 59–60, 65, 78–80,
    80, 93, 94, 104, 105, 115, 181
  of gorillas, 104, 105–106, 162
  of orangutans, 195–196
  of parrots, 11, 28, 245
Vocal cords
  of chimpanzees, 11, 12
  of humans, 11–12
Vocal development, in nonhuman primates,
  12–13

Vocalizations
  by bonobos, 14, 17, 178–179
  by chimpanzees, 54, 58–59, 60, 62, 80
    emotional arousal-related, 13–14, 75
    involuntary nature of, 13–14, 75
  by dolphins, 214, 217, 221, 233
  by elephants, 26
  by orangutans, 54
  by parrots, 239–240
    during model/rival training, 249
  relationship with manual dexterity, 14, 15
Vocal prostheses, for apes, 179–180, 275–276
Vocal tract
  of apes, prosthetic modification of, 179–180,
    275–276
  of chimpanzees, 11, 12
  of parrots, 11, 15
Von Glaserfeld, Ernst, 128, 129–130, 131,
  146–147, 148, 262
Von Osten, Wilhelm, 1, 2, 47, 50–53
Von Uexkuell, Jakob, 203
Vowel sounds, 12

Wallman, Joe, 259–260, 261–265
Warner, Harold, 127, 128, 145
Warner Brothers Studio, 89
Warning calls, 18, 25, 276–277
Warning function, of animal communication,
  26
Washburn, David, 4
Washburn, Margaret Floy, 21–22, 123
Washoe (chimpanzee), 5, 48, 63, 65, 66, 69–85,
  78, 87, 90, 91–92, 126, 154–155, 171
  biting behavior of, 31, 92–93
  offspring of, 6, 88, 92–93
  sign language use by, 4, 37, 69–73, 75, 104,
    115, 155
    deaf researchers' interpretation of,
      257–258, 260
    effect of enculturation on, 156–157
    grammatical category understanding in,
      80–82
    grounding in, 40–41
    "natural' signs in, 27
    properties of, 76–78
    vocabulary in, 70, 78–80, 115
Washoe Project, 72, 73, 126, 127. *See also*
  Washoe (chimpanzee)
"Wh-questions," 80–82, 117
Why Chimpanzee Can Read (Premack), 114
Wild Minds (Hauser), 121
William of Occam, 1, 33
Wise, Steve, 99, 100, 102
Wittgenstein, Ludwig, 217
Woodruff, G., 52